Maximum   SC = Sufficient Condition   NC = Necessary Condition

$f(x^*)$ ND: SC for local max @ x^*.
$f(x^*)$ NSD: NC for local max @ x^*.
$f(x)$ ND $\forall x \in \mathbb{R}^n$: SC for global max @ x
$f(x)$ NSD $\forall x \in \mathbb{R}^n$: NC for global max @ X

Minimum
$f(x^*)$ PD: SC for local min @ x^*.
$f(x^*)$ PSD: NC for local min @ x^*.
$f(x)$ PD $\forall x \in \mathbb{R}^n$: SC for global min @ x. (f is strictly convex),
$f(x)$ PSD $\forall x \in \mathbb{R}^n$: NC for global min @ x. (f is convex).

Indefinite: $D^2 f(x^*)$ is indefinite $\Rightarrow$ A "saddle point."

Matrix Examples

$\rightarrow \begin{pmatrix} 1 & 0 & 0 \\ 0 & 1 & 0 \\ 0 & 0 & 1 \end{pmatrix}$ NSD $\rightarrow \begin{pmatrix} -1 & 0 & 0 \\ 0 & -1 & 0 \\ 0 & 0 & 0 \end{pmatrix}$ ND $\Rightarrow \begin{pmatrix} -1 & 0 & 0 \\ 0 & -1 & 0 \\ 0 & 0 & -1 \end{pmatrix}$ Indefinite $\rightarrow \begin{pmatrix} 1 & 0 & 0 \\ 0 & -1 & 0 \\ 0 & 0 & 0 \end{pmatrix}$

Matrix info!

matrices A, B, & C, if A(B+C) is well defined, then A(B-C) = AB - AC.
real matrices A & B, if AB is well defined, then $(AB)^T = B^T A^T$
any matrix $A_{n \times n}$ we always have $|A^T| = |A|$,
symmetric matrix $A_{n \times n}$, the quadratic form $x^T A x$ is NSD if $x^T A x \leq 0$
all $x \in \mathbb{R}^n$.

addition is commutative, subtraction is not. Mult is not commutative.
$A + B = B + A$ addition & Mult are assoc & distributive.
scalar Mult.   Matrix Mult $\rightarrow$ Ex. $\begin{bmatrix} 1 & 2 & 3 \\ 4 & 5 & 6 \end{bmatrix} \times \begin{bmatrix} 0 & 1 \\ 2 & 4 \\ 3 & 5 \end{bmatrix} = \begin{bmatrix} 13 & 24 \\ 28 & 54 \end{bmatrix}$

$A_{m \times n}$, $t \in \mathbb{R}$ $A \cdot B = C$
$tA = [t a_{ij}]$ $m \times n \; n \times p = m \times p$ 2×3 3×2 2×2
$2 \begin{bmatrix} 1 & 2 & 3 \\ 4 & 5 & 6 \end{bmatrix} = \begin{bmatrix} 2 & 4 & 6 \\ 8 & 10 & 12 \end{bmatrix}$

                              Identity matrix
                              $I_n = \begin{bmatrix} 1 & 0 \\ 0 & 1 \end{bmatrix}$
Transpose:
$A = \begin{bmatrix} 1 & 2 \\ 3 & 4 \\ 5 & 6 \end{bmatrix}$ $A^T = \begin{bmatrix} 1 & 3 & 5 \\ 2 & 4 & 6 \end{bmatrix}$ Properties:
$(A^T)^T = A$
$(A+B)^T = A^T + B^T$
$(AB)^T = B^T A^T$

Row Vector:
$x = (1,2) \quad y = (3,4)$
$y^t = \begin{pmatrix} 1 \\ 2 \end{pmatrix} (3,4) = \begin{bmatrix} 3 & 4 \\ 6 & 8 \end{bmatrix}$
$2 \times 1 \quad 1 \times 2 \quad 2 \times 2$

$x^t = \begin{pmatrix} 3 \\ 4 \end{pmatrix} (1 2) = \begin{bmatrix} 3 & 6 \\ 4 & 8 \end{bmatrix}$

$x y = (1 2) \begin{pmatrix} 3 \\ 4 \end{pmatrix} = [11]$ "Dot Product"
$1 \times 2 \; 2 \times 1 \quad 1 \times 1$

matrix has inverse, it is regular. No inverse, then nonregular, A^{-1}
EX. $A = \begin{pmatrix} 2 & 1 \\ 1 & 1 \end{pmatrix}$, $B = \begin{pmatrix} 1 & -1 \\ -1 & 2 \end{pmatrix}$

$AB = \begin{bmatrix} 2(1)+(1)(-1) & 2(-1)+1(2) \\ (1)(1)+(1)(-1) & 1(-1)+1(2) \end{bmatrix} = \begin{bmatrix} 1 & 0 \\ 0 & 1 \end{bmatrix} = I_2$

rank of a matrix
  # of independent rows
  # of independent columns.

null matrix is PSD & NSD

Monotonic function

Let $A \subseteq \mathbb{R}^N$, $f: A \to \mathbb{R}$

$x, y \in \mathbb{R}^N$,

fix

a) Weakly monotonically increasing

if $x \leq y \Rightarrow f(x) \geq f(y)$

b) strictly monotonically increasing

if $x < y \Rightarrow f(x) > f(y)$

For every monotonic function, there is always an inverse function.

Ex, $y = \frac{1}{3}x$

reverse: $x = \frac{1}{3}y$

$3x = y$

$y = 3x$

Transform

$T: A \to B$

Let $A = B = \{f: \mathbb{R} \to \mathbb{R}\}$

Let $T(f) = T \circ f = \ln(f)$

consider $f(x) = 2x$

Then where $g: \mathbb{R} \to \mathbb{R}$ where $g = T \circ f$

$g(x) = T(f(x)) = T(2x) = \ln(2x) = \ln(2) + \ln(x)$

Convex / Concave

Function in Q concave if all upper contour sets are convex.

"   "   " Strictly "   "   "   "   "   "   "   " strictly convex.

"   " Q convex if "   "   "   "   "   "   "   " convex

"   " strictly " if "   "   "   "   "   " are strictly ".

all concave functions are Q-concave.

all strictly concave functions are strictly Q concave.

Same for convex.

Ex, $f(\ell) = e^{-\ell+1}$

$f'(\ell) = -1(e^{-\ell+1})$

Ex, $f(x) = e^{(3x^2-2x)}$

$f'(x) = (6x-2) e^{3x^2-2x}$

Ex, $f(u) = 3^u$

$f'(u) = 3^u \ln 3$

Ex, $f(x) = 3^{x^2}$

$f'(x) = 3^{x^2} \ln 3 (2x)$

$f: \mathbb{R}^2 \to \mathbb{R}^3$ — cannot do gradient

$\mathbb{R}^2 \to \mathbb{R}$

$g(x,y) = x \ln(y)$

$Dg(x) = \begin{bmatrix} \ln(y) & x/y \end{bmatrix}$

$\nabla g(x) = \begin{pmatrix} \ln(y) \\ \frac{x}{y} \end{pmatrix}$

$\nabla g(x) = \begin{bmatrix} Dg(x) \end{bmatrix}^{\top}$

Economists' Mathematical Manual

Vector addition

$$V_a + V_b = (N_1^a + x_1^b,\ N_2^a + x_2^b)$$

...ometric Interpretation of Determinant

$$w/\ A = \begin{bmatrix} a & b \\ c & d \end{bmatrix} \quad V_1 = \begin{bmatrix} A \\ c \end{bmatrix} \quad V_2 = \begin{bmatrix} b \\ d \end{bmatrix}$$

$|A| = $ area

1) $A = \begin{bmatrix} 3 & 0 \\ 0 & 2 \end{bmatrix}$

$| = 6$

A vec; 6

$A = \begin{bmatrix} 9 & 6 \\ 6 & 4 \end{bmatrix}$

area $= 0$

$|A| = 36 - 36$

0

$1 = \begin{pmatrix} -1 & 3 \\ 2 & 0 \end{pmatrix}$ $(A - r I_n) = \begin{pmatrix} -1-r & 3 \\ 2 & -r \end{pmatrix} = ((-1-r)\cdot -r) - 6 = (r+3)(r-2)$

$r = -3 \quad r = 2$

find char. vector

$A - r I_2) V = 0$

$= \begin{pmatrix} -1+3 & 3 \\ 2 & 3 \end{pmatrix} \begin{pmatrix} v_1 \\ v_2 \end{pmatrix} = 0$

$\begin{pmatrix} 2 & 3 \\ 2 & 3 \end{pmatrix} \begin{pmatrix} v_1 \\ v_2 \end{pmatrix} = 0$

$2 v_1 + 3 v_2 = 0$

$2 v_1 + 3 v_2 = 0$

$v_2 = -2/3\, v_1$

$\begin{pmatrix} 1 \\ -2/3 \end{pmatrix}, \begin{pmatrix} 3 \\ -2 \end{pmatrix}$

normalize so vector length is 1 $\cdot \|v\| = 1$

$\sqrt{2^2 + 3^2} = \sqrt{13}$ so $V_a = \begin{pmatrix} \frac{3}{\sqrt{13}} \\ \frac{-2}{\sqrt{13}} \end{pmatrix}$

$v_b = 2 \begin{pmatrix} -1-2 & 3 \\ 2 & -2 \end{pmatrix} \begin{pmatrix} v_1 \\ v_2 \end{pmatrix}$

$= \begin{pmatrix} -3 & 3 \\ 3 & -2 \end{pmatrix} \begin{pmatrix} v_1 \\ v_2 \end{pmatrix}$

$-3 v_1 + 3 v_2 = 0$

$3 v_1 - 2 v_2 = 0$

$v_2 = v_1$

$\begin{pmatrix} 1 \\ 1 \end{pmatrix}$ $\sqrt{1^2 + 1^2} = \sqrt{2}$ $V_b = \begin{pmatrix} \frac{1}{\sqrt{2}} \\ \frac{1}{\sqrt{2}} \end{pmatrix}$

Finding rank of matrix

Row echelon form
1) all zero rows are below others
2) Leading element of any row in to the right of the leading element of all rows above it.
3) all elements in col. below a leading element are zeros.

Ex. $\begin{bmatrix} 1 & \frac{1}{4} & 0 \\ 0 & 5\frac{1}{2} & 11 \\ 0 & 0 & 18 \end{bmatrix}$

Reduced row echelon form
4) all leading elements are one.
5) each leading element only non-zero element in column.

Ex. $\begin{bmatrix} 1 & 0 & 0 \\ 0 & 1 & 0 \\ 0 & 0 & 1 \end{bmatrix}$

Cramer's Rule

Ex. $5x_1 + 3x_2 = 30$
$6x_1 - 2x_2 = 8$

$|A| = \begin{vmatrix} 5 & 3 \\ 6 & -2 \end{vmatrix} = -28$

$|A_1| = \begin{vmatrix} 30 & 3 \\ 8 & -2 \end{vmatrix} = -84$

$|A_2| = \begin{vmatrix} 5 & 30 \\ 6 & 8 \end{vmatrix} = -140$

$x_1^* = \dfrac{|A_1|}{|A|} = \dfrac{-84}{-28} = 3$

$x_2^* = \dfrac{|A_2|}{|A|} = \dfrac{-140}{-28} = 5$

L'Hospital's Rule : Take derivative if denominator = 0

$Q(x) = ax^2 = xax$
PD iff $a > 0$
ND iff $a < 0$

Knut Sydsæter · Arne Strøm · Peter Berck

Economists' Mathematical Manual

$f'(x)$: local slope
$f''(x)$: Rate of Δ of slope

$y = f(x)$

Fourth Edition

$\leftarrow f(x) \geq 0 \rightarrow$
$f(x) \leq 0$
$\leftarrow f(x) \geq 0 \rightarrow x$

$\leftarrow f'(x) \geq 0 \rightarrow$
$\leftarrow f'(x) \leq 0 \rightarrow$ $\leftarrow f'(x) \geq 0 \rightarrow$ $\leftarrow f'(x) \leq 0 \rightarrow$

$\leftarrow f''(x) \leq 0 \rightarrow$

$\leftarrow f''(x) \geq 0 \rightarrow$ $\leftarrow f''(x) \leq 0 \rightarrow$

$f(x) = \ln(x)$
$f'(x) = \frac{1}{x} = x^{-1}$
$f''(x) = -x^{-2} = \frac{-1}{x^2}$

Ex, $f(x) = x^2 - 6x + 11$
$f'(x) = 2x - 6$
$= 0 \quad 2x = 6$
$\quad x = 3$
$f''(x) = 2 > 0$
min @ $x = 3$

$2 +$ ∪
3

EX. $f(x) = x^3 - 6x^2 + 12x - 5$
$f'(x) = 3x^2 - 12x + 12$
$= 0 \quad (x = 2$
$f''(x) = 6x - 12$
$6(2) - 12 = 0$

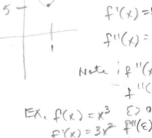

Ex. $f(x) = x^4 - 4x^3 - 6x^2 - 4x + 5$
$f'(x) = 4x^3 - 12x^2 + 12x - 4$
$= 0 \quad (x-1)^3 = 0$
$x = 1$

$5 +$ ∪

EX. $f(x) = x^4$
$f'(x) = 4x^3$
$f''(x) = 12x^2 \, \varepsilon > 0$
$f''(\varepsilon) > 0$
$f''(-\varepsilon) > 0$

$y = x^4$
$f''(x) > 0$ $f''(x) > 0$
$f''(x) = 0$

Note if $f''(x) > 0$ convex
$f''(x) < 0$ concave

  Springer

$f''(x) = 12x^2 - 24x - 12$
$12(1) - 24(1) - 12$
$= 0$

EX. $f(x) = x^3 \quad \varepsilon > 0$
$f'(x) = 3x^2 \quad f''(\varepsilon) = 6\varepsilon > 0$
$f''(x) = 6x \quad f''(-\varepsilon) = -6\varepsilon < 0$
cannot be extremum

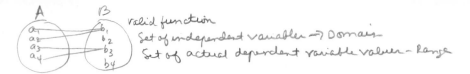

valid function
Set of independent variables → Domain
Set of actual dependent variable values – Range

Professor Knut Sydsæter
University of Oslo
Department of Economics
P.O. Box 10955 Blindern
NO-0317 Oslo
Norway
knutsy@econ.uio.no

Associate Professor Arne Strøm
University of Oslo
Department of Economics
P.O. Box 10955 Blindern
NO-0317 Oslo
Norway
arne.strom@econ.uio.no

Professor Peter Berck
University of California, Berkeley
Department of Agricultural and
Resource Economics
Berkeley, CA 94720-3310
USA
pberck@berkeley.edu

Homogeneous Functions!

$$f: \mathbb{R}^2 \to \mathbb{R}, \; f(x, x_2) = 2x_1 + (x_2)^2$$
$$f(tx) = 2(tx_1) + (tx_2)^2 = 2t x_1 + t^2 x_2^2$$
$$= t(2 x_1 + t x_2^2)$$

Cannot be stated in the form $t^n f(x)$, so Not homogeneous.

$$f: \mathbb{R}^2 \to \mathbb{R}, \; f(x, x_2) = \pi \circ$$
$$f(t x) = \pi = t\pi$$
$$= t^0 f(x) \quad HOD - 0$$

ISBN 978-3-540-26088-2   e-ISBN 978-3-540-28518-2
DOI 10.1007/978-3-540-28518-2
Springer Heidelberg Dordrecht London New York

Library of Congress Control Number: 2009937018

Cover design: WMXDesign GmbH, Heidelberg

Printed on acid-free paper

Springer is part of Springer Science+Business Media (www.springer.com)

Preface to the fourth edition

The fourth edition is augmented by more than 70 new formulas. In particular, we have included some key concepts and results from trade theory, games of incomplete information and combinatorics. In addition there are scattered additions of new formulas in many chapters.

Again we are indebted to a number of people who has suggested corrections, improvements and new formulas. In particular, we would like to thank Jens-Henrik Madsen, Larry Karp, Harald Goldstein, and Geir Asheim.

In a reference book, errors are particularly destructive. We hope that readers who find our remaining errors will call them to our attention so that we may purge them from future editions.

Oslo and Berkeley, May 2005

Knut Sydsæter, Arne Strøm, Peter Berck

From the preface to the third edition

The practice of economics requires a wide-ranging knowledge of formulas from mathematics, statistics, and mathematical economics. With this volume we hope to present a formulary tailored to the needs of students and working professionals in economics. In addition to a selection of mathematical and statistical formulas often used by economists, this volume contains many purely economic results and theorems. It contains just the formulas and the minimum commentary needed to relearn the mathematics involved. We have endeavored to state theorems at the level of generality economists might find useful. In contrast to the economic maxim, "everything is twice more continuously differentiable than it needs to be", we have usually listed the regularity conditions for theorems to be true. We hope that we have achieved a level of explication that is accurate and useful without being pedantic.

During the work with this book we have had help from a large group of people. It grew out of a collection of mathematical formulas for economists originally compiled by Professor B. Thalberg and used for many years by Scandinavian students and economists. The subsequent editions were much improved by the suggestions and corrections of: G. Asheim, T. Akram, E. Biørn, T. Ellingsen, P. Frenger, I. Frihagen, H. Goldstein, F. Greulich, P. Hammond, U. Hassler, J. Heldal, Aa. Hylland, G. Judge, D. Lund, M. Machina, H. Mehlum, K. Moene, G. Nordén, A. Rødseth, T. Schweder, A. Seierstad, L. Simon, and B. Øksendal.

As for the present third edition, we want to thank in particular, Olav Bjerkholt, Jens-Henrik Madsen, and the translator to Japanese, Tan-no Tadanobu, for very useful suggestions.

Oslo and Berkeley, November 1998

Knut Sydsæter, Arne Strøm, Peter Berck

Contents

quadratic mean. Slutsky's theorem. Limiting distribution. Consistency. Testing. Power of a test. Type I and type II errors. Level of significance. Significance probability (P-value). Weak and strong law of large numbers. Central limit theorem.

Chapter 1

Set Theory. Relations. Functions

1.1 $x \in A, \quad x \notin B$

The element x belongs to the set A, but x does not belong to the set B.

1.2 $A \subset B \iff$ Each element of A is also an element of B.

A is a *subset* of B. Often written $A \subseteq B$.

1.3   If S is a set, then the set of all elements x in S with property $\varphi(x)$ is written

$$A = \{x \in S : \varphi(x)\}$$

If the set S is understood from the context, one often uses a simpler notation:

$$A = \{x : \varphi(x)\}$$

General notation for the specification of a set. For example,
$\{x \in \mathbb{R} : -2 \leq x \leq 4\} = [-2, 4]$.

1.4   The following logical operators are often used when P and Q are statements:

- $P \wedge Q$ means "P and Q"
- $P \vee Q$ means "P or Q"
- $P \Rightarrow Q$ means "if P then Q" (or "P only if Q", or "P implies Q")
- $P \Leftarrow Q$ means "if Q then P"
- $P \Leftrightarrow Q$ means "P if and only if Q"
- $\neg P$ means "not P"

Logical operators. (Note that "P or Q" means "either P or Q or both".)

1.5

P	Q	$\neg P$	$P \wedge Q$	$P \vee Q$	$P \Rightarrow Q$	$P \Leftrightarrow Q$
T	T	F	T	T	T	T
T	F	F	F	T	F	F
F	T	T	F	T	T	F
F	F	T	F	F	T	T

Truth table for logical operators. Here T means "true" and F means "false".

1.6

- P is a *sufficient condition* for Q: $P \Rightarrow Q$
- Q is a *necessary condition* for P: $P \Rightarrow Q$
- P is a *necessary and sufficient condition* for Q: $P \Leftrightarrow Q$

Frequently used terminology.

$$A \cup B = \{x : x \in A \vee x \in B\} \quad (A \; union \; B)$$
$$A \cap B = \{x : x \in A \wedge x \in B\} \quad (A \; intersection \; B)$$
$$A \setminus B = \{x : x \in A \wedge x \notin B\} \quad (A \; minus \; B)$$

1.7 $\quad A \triangle B = (A \setminus B) \cup (B \setminus A) \quad (symmetric \; difference)$

If all the sets in question are contained in some "universal" set Ω, one often writes $\Omega \setminus A$ as

$$A^c = \{x : x \notin A\} \quad (\text{the } complement \text{ of } A)$$

Basic set operations. $A \setminus B$ is called the *difference* between A and B. An alternative symbol for A^c is $\complement A$.

  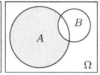

$\qquad A \cup B \qquad\qquad A \cap B \qquad\qquad A \setminus B \qquad\qquad A^c \qquad\qquad A \triangle B$

$$A \cap (B \cup C) = (A \cap B) \cup (A \cap C)$$
$$A \cup (B \cap C) = (A \cup B) \cap (A \cup C)$$
$$A \triangle B = (A \cup B) \setminus (A \cap B)$$

1.8 $\quad (A \triangle B) \triangle C = A \triangle (B \triangle C)$

$$A \setminus (B \cup C) = (A \setminus B) \cap (A \setminus C)$$
$$A \setminus (B \cap C) = (A \setminus B) \cup (A \setminus C)$$
$$A \cup B)^c = A^c \cap B^c$$
$$(A \cap B)^c = A^c \cup B^c$$

Important identities in set theory. The last four identities are called *De Morgan's laws.*

1.9 $\quad A_1 \times A_2 \times \cdots \times A_n =$
$\qquad \{(a_1, a_2, \ldots, a_n) : a_i \in A_i \text{ for } i = 1, 2, \ldots, n\}$

The *Cartesian product* of the sets $A_1, A_2, \ldots, A_n$.

1.10 $\quad R \subset A \times B$

Any subset R of $A \times B$ is called a *relation* from the set A into the set B.

1.11 $\quad \begin{aligned} xRy &\iff (x,y) \in R \\ x\not\!Ry &\iff (x,y) \notin R \end{aligned}$

Alternative notations for a relation and its negation. We say that x is in R-relation to y if $(x,y) \in R$.

1.12
- $\text{dom}(R) \; = \{a \in A : (a,b) \in R \text{ for some } b \text{ in } B\}$
  $\qquad\qquad = \{a \in A : aRb \text{ for some } b \text{ in } B\}$
- $\text{range}(R) = \{b \in B : (a,b) \in R \text{ for some } a \text{ in } A\}$
  $\qquad\qquad = \{b \in B : aRb \text{ for some } a \text{ in } A\}$

The *domain* and *range* of a relation.

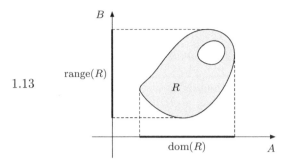

1.13

Illustration of the domain and range of a relation, R, as defined in (1.12). The shaded set is the *graph* of the relation.

1.14 $\quad R^{-1} = \{(b,a) \in B \times A : (a,b) \in R\}$

The *inverse* relation of a relation R from A to B. R^{-1} is a relation from B to A.

1.15 Let R be a relation from A to B and S a relation from B to C. Then we define the *composition* $S \circ R$ of R and S as the set of all (a,c) in $A \times C$ such that there is an element b in B with aRb and bSc. $S \circ R$ is a relation from A to C.

$S \circ R$ is the composition of the relations R and S.

A relation R from A to A itself is called a *binary relation* in A. A binary relation R in A is said to be

- *reflexive* if aRa for every a in A;
- *irreflexive* if $a\not\!Ra$ for every a in A;

1.16
- *complete* if aRb or bRa for every a and b in A with $a \neq b$;

Special relations.

- *transitive* if aRb and bRc imply aRc;
- *symmetric* if aRb implies bRa;
- *antisymmetric* if aRb and bRa implies $a = b$;
- *asymmetric* if aRb implies $b\not\!Ra$.

A binary relation R in A is called

- a *preordering* (or a *quasi-ordering*) if it is reflexive and transitive;
- a *weak ordering* if it is transitive and complete;

1.17
- a *partial ordering* if it is reflexive, transitive, and antisymmetric;
- a *linear* (or *total*) *ordering* if it is reflexive, transitive, antisymmetric, and complete;
- an *equivalence relation* if it is reflexive, transitive, and symmetric.

Special relations. (The terminology is not universal.) Note that a linear ordering is the same as a partial ordering that is also complete.

Order relations are often denoted by symbols like $\preccurlyeq, \leq, \ll$, etc. The inverse relations are then denoted by $\succcurlyeq, \geq, \gg$, etc.

- The relation $=$ between real numbers is an equivalence relation.
- The relation $\leq$ between real numbers is a linear ordering.
- The relation $<$ between real numbers is a weak ordering that is also irreflexive and asymmetric.
- The relation $\subset$ between subsets of a given set is a partial ordering.

1.18
- The relation $\mathbf{x} \preceq \mathbf{y}$ ($\mathbf{y}$ is *at least as good as* $\mathbf{x}$) in a set of commodity vectors is usually assumed to be a complete preordering.
- The relation $\mathbf{x} \prec \mathbf{y}$ ($\mathbf{y}$ is *(strictly) preferred to* $\mathbf{x}$) in a set of commodity vectors is usually assumed to be irreflexive, transitive, (and consequently asymmetric).
- The relation $\mathbf{x} \sim \mathbf{y}$ ($\mathbf{x}$ is *indifferent to* $\mathbf{y}$) in a set of commodity vectors is usually assumed to be an equivalence relation.

Examples of relations. For the relations $\mathbf{x} \preceq \mathbf{y}$, $\mathbf{x} \prec \mathbf{y}$, and $\mathbf{x} \sim \mathbf{y}$, see Chap. 26.

1.19 Let $\preccurlyeq$ be a preordering in a set A. An element g in A is called a *greatest element* for $\preccurlyeq$ in A if $x \preccurlyeq g$ for every x in A. An element m in A is called a *maximal element* for $\preccurlyeq$ in A if $x \in A$ and $m \preccurlyeq x$ implies $x \preccurlyeq m$. A *least element* and a *minimal element* for $\preccurlyeq$ are a greatest element and a maximal element, respectively, for the inverse relation $\succcurlyeq$ of $\preccurlyeq$.

The definition of a greatest element, a maximal element, a least element, and a minimal element of a preordered set.

1.20 If $\preccurlyeq$ is a preordering in A and M is a subset of A, an element b in A is called an *upper bound* for M (w.r.t. $\preccurlyeq$) if $x \preccurlyeq b$ for every x in M. A *lower bound* for M is an element a in A such that $a \preccurlyeq x$ for all x in M.

Definition of upper and lower bounds.

1.21 If $\preccurlyeq$ is a preordering in a nonempty set A and if each linearly ordered subset M of A has an upper bound in A, then there exists a maximal element for $\preccurlyeq$ in A.

Zorn's lemma. (Usually stated for partial orderings, but also valid for preorderings.)

1.22 A relation R from A to B is called a *function* or *mapping* if for every a in A, there is a unique b in B with aRb. If the function is denoted by f, then we write $f(a) = b$ for afb, and the *graph* of f is defined as:

$$\text{graph}(f) = \{(a,b) \in A \times B : f(a) = b\}.$$

The definition of a function and its graph.

1.23 A function f from A to B ($f : A \to B$) is called

- *injective* (or *one-to-one*) if $f(x) = f(y)$ implies $x = y$;
- *surjective* (or *onto*) if $\text{range}(f) = B$;
- *bijective* if it is injective and surjective.

Important concepts related to functions.

1.24 If $f : A \to B$ is bijective (i.e. both one-to-one and onto), it has an *inverse function* $g : B \to A$, defined by $g(f(u)) = u$ for all u in A.

Characterization of inverse functions. The inverse function of f is often denoted by f^{-1}.

1.25

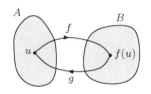

Illustration of the concept of an inverse function.

1.26 If f is a function from A to B, and $C \subset A$, $D \subset B$, then we use the notation

- $f(C) = \{f(x) : x \in C\}$
- $f^{-1}(D) = \{x \in A : f(x) \in D\}$

$f(C)$ is called the *image* of A under f, and $f^{-1}(D)$ is called the *inverse image* of D.

1.27 If f is a function from A to B, and $S \subset A$, $T \subset A$, $U \subset B$, $V \subset B$, then

- $f(S \cup T) = f(S) \cup f(T)$
- $f(S \cap T) \subset f(S) \cap f(T)$
- $f^{-1}(U \cup V) = f^{-1}(U) \cup f^{-1}(V)$
- $f^{-1}(U \cap V) = f^{-1}(U) \cap f^{-1}(V)$
- $f^{-1}(U \setminus V) = f^{-1}(U) \setminus f^{-1}(V)$

Important facts. The inclusion $\subset$ in
$$f(S \cap T) \subset f(S) \cap f(T)$$
cannot be replaced by $=$.

1.28 Let $\mathbb{N} = \{1, 2, 3, \ldots\}$ be the set of natural numbers, and let $\mathbb{N}_n = \{1, 2, 3, \ldots, n\}$. Then:

- A set A is *finite* if it is empty, or if there exists a one-to-one function from A onto $\mathbb{N}_n$ for some natural number n.
- A set A is *countably infinite* if there exists a one-to-one function of A onto $\mathbb{N}$.

A set that is either finite or countably infinite, is often called *countable*. The set of rational numbers is countably infinite, while the set of real numbers is not countable.

Suppose that $A(n)$ is a statement for every natural number n and that

1.29

- $A(1)$ is true,
- if the induction hypothesis $A(k)$ is true, then $A(k+1)$ is true for each natural number k.

Then $A(n)$ is true for all natural numbers n.

The *principle of mathematical induction.*

References

See Halmos (1974), Ellickson (1993), and Hildenbrand (1974).

Steps to find inverse of function.
① Replace $f(x)$ w/ y.
② Switch x's & y's
③ Solve for y
④ Replace y w/ $f^{-1}(x)$

EX, $f(x) = \sqrt{x+4} \quad -3$

$y = \sqrt{x+4} -3$

$x = \sqrt{y+4} -3$

$3+x = \sqrt{y+4}$

$(3+x)^2 = (\sqrt{y+4})^2$

$x^2 + 6x + 9 = y+4$

$x^2 + 6x + 5 = y$

$f^{-1}(x) = x^2 + 6x + 5$

Chapter 2

Equations. Functions of one variable. Complex numbers

2.1 $\quad ax^2 + bx + c = 0 \iff x_{1,2} = \dfrac{-b \pm \sqrt{b^2 - 4ac}}{2a}$

The roots of the general *quadratic* equation. They are real provided $b^2 \geq 4ac$ (assuming that a, b, and c are real).

2.2 If x_1 and x_2 are the roots of $x^2 + px + q = 0$, then
$$x_1 + x_2 = -p, \qquad x_1 x_2 = q$$

Viète's rule.

2.3 $\quad ax^3 + bx^2 + cx + d = 0$

The general *cubic* equation.

2.4 $\quad x^3 + px + q = 0$

(2.3) reduces to the form (2.4) if x in (2.3) is replaced by $x - b/3a$.

2.5 $x^3 + px + q = 0$ with $\Delta = 4p^3 + 27q^2$ has

- three different real roots if $\Delta < 0$;
- three real roots, at least two of which are equal, if $\Delta = 0$;
- one real and two complex roots if $\Delta > 0$.

Classification of the roots of (2.4) (assuming that p and q are real).

2.6 The solutions of $x^3 + px + q = 0$ are
$x_1 = u + v$, $x_2 = \omega u + \omega^2 v$, and $x_3 = \omega^2 u + \omega v$,
where $\omega = -\frac{1}{2} + \frac{i}{2}\sqrt{3}$, and

$$u = \sqrt[3]{-\frac{q}{2} + \frac{1}{2}\sqrt{\frac{4p^3 + 27q^2}{27}}}$$

$$v = \sqrt[3]{-\frac{q}{2} - \frac{1}{2}\sqrt{\frac{4p^3 + 27q^2}{27}}}$$

Cardano's formulas for the roots of a cubic equation. i is the imaginary unit (see (2.75)) and ω is a complex third root of 1 (see (2.88)). (If complex numbers become involved, the cube roots must be chosen so that $3uv = -p$. Don't try to use these formulas unless you have to!)

If x_1, x_2, and x_3 are the roots of the equation $x^3 + px^2 + qx + r = 0$, then

2.7
$$x_1 + x_2 + x_3 = -p$$
$$x_1x_2 + x_1x_3 + x_2x_3 = q$$
$$x_1x_2x_3 = -r$$

Useful relations.

2.8 $\quad P(x) = a_n x^n + a_{n-1} x^{n-1} + \cdots + a_1 x + a_0$

A *polynomial* of degree n. $(a_n \neq 0.)$

2.9 For the polynomial $P(x)$ in (2.8) there exist constants x_1, x_2, ..., x_n (real or complex) such that
$$P(x) = a_n(x - x_1) \cdots (x - x_n)$$

The *fundamental theorem of algebra*. x_1, ..., x_n are called *zeros* of $P(x)$ and *roots* of $P(x) = 0$.

2.10
$$x_1 + x_2 + \cdots + x_n = -\frac{a_{n-1}}{a_n}$$
$$x_1x_2 + x_1x_3 + \cdots + x_{n-1}x_n = \sum_{i<j} x_i x_j = \frac{a_{n-2}}{a_n}$$
$$x_1x_2 \cdots x_n = (-1)^n \frac{a_0}{a_n}$$

Relations between the roots and the coefficients of $P(x) = 0$, where $P(x)$ is defined in (2.8). (Generalizes (2.2) and (2.7).)

2.11 If a_{n-1}, ..., a_1, a_0 are all integers, then any integer root of the equation
$$x^n + a_{n-1} x^{n-1} + \cdots + a_1 x + a_0 = 0$$
must divide a_0.

Any integer solutions of $x^3 + 6x^2 - x - 6 = 0$ must divide -6. (In this case the roots are ± 1 and -6.)

2.12 Let k be the number of changes of sign in the sequence of coefficients a_n, a_{n-1}, ..., a_1, a_0 in (2.8). The number of positive real roots of $P(x) = 0$, counting the multiplicities of the roots, is k or k minus a positive even number. If $k = 1$, the equation has exactly one positive real root.

Descartes's rule of signs.

2.13 The graph of the equation
$$Ax^2 + Bxy + Cy^2 + Dx + Ey + F = 0$$
is
• an ellipse, a point or empty if $4AC > B^2$;
• a parabola, a line, two parallel lines, or empty if $4AC = B^2$;
• a hyperbola or two intersecting lines if $4AC < B^2$.

Classification of *conics*. A, B, C not all 0.

2.14	$x = x'\cos\theta - y'\sin\theta, \quad y = x'\sin\theta + y'\cos\theta$ with $\cot 2\theta = (A - C)/B$	Transforms the equation in (2.13) into a quadratic equation in x' and y', where the coefficient of $x'y'$ is 0.
2.15	$d = \sqrt{(x_2 - x_1)^2 + (y_2 - y_1)^2}$	The (Euclidean) *distance* between the points (x_1, y_1) and (x_2, y_2).
2.16	$(x - x_0)^2 + (y - y_0)^2 = r^2$	*Circle* with center at (x_0, y_0) and radius r.
2.17	$\dfrac{(x - x_0)^2}{a^2} + \dfrac{(y - y_0)^2}{b^2} = 1$	*Ellipse* with center at (x_0, y_0) and axes parallel to the coordinate axes.
2.18	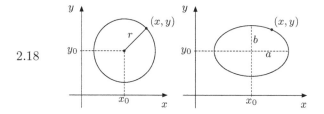	Graphs of (2.16) and (2.17).
2.19	$\dfrac{(x - x_0)^2}{a^2} - \dfrac{(y - y_0)^2}{b^2} = \pm 1$	*Hyperbola* with center at (x_0, y_0) and axes parallel to the coordinate axes.
2.20	Asymptotes for (2.19): $y - y_0 = \pm\dfrac{b}{a}(x - x_0)$	Formulas for asymptotes of the hyperbolas in (2.19).
2.21	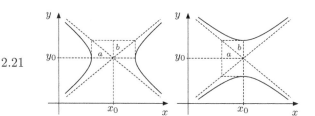	Hyperbolas with asymptotes, illustrating (2.19) and (2.20), corresponding to $+$ and $-$ in (2.19), respectively. The two hyperbolas have the same asymptotes.
2.22	$y - y_0 = a(x - x_0)^2, \quad a \neq 0$	*Parabola* with vertex (x_0, y_0) and axis parallel to the y-axis.
2.23	$x - x_0 = a(y - y_0)^2, \quad a \neq 0$	*Parabola* with vertex (x_0, y_0) and axis parallel to the x-axis.

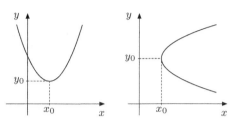

Parabolas illustrating (2.22) and (2.23) with $a > 0$.

A function f is

- *increasing* if
$$x_1 < x_2 \Rightarrow f(x_1) \le f(x_2)$$
- *strictly increasing* if
$$x_1 < x_2 \Rightarrow f(x_1) < f(x_2)$$
- *decreasing* if
$$x_1 < x_2 \Rightarrow f(x_1) \ge f(x_2)$$
- *strictly decreasing* if
$$x_1 < x_2 \Rightarrow f(x_1) > f(x_2)$$

2.25

- *even* if $f(x) = f(-x)$ for all x
- *odd* if $f(x) = -f(-x)$ for all x
- *symmetric about the line* $x = a$ if
$$f(a + x) = f(a - x) \text{ for all } x$$
- *symmetric about the point* $(a, 0)$ if
$$f(a - x) = -f(a + x) \text{ for all } x$$
- *periodic* (with period k) if there exists a number $k > 0$ such that
$$f(x + k) = f(x) \text{ for all } x$$

Properties of functions.

- If $y = f(x)$ is replaced by $y = f(x) + c$, the graph is moved upwards by c units if $c > 0$ (downwards if c is negative).
- If $y = f(x)$ is replaced by $y = f(x + c)$, the graph is moved c units to the left if $c > 0$ (to the right if c is negative).

2.26

- If $y = f(x)$ is replaced by $y = cf(x)$, the graph is stretched vertically if $c > 0$ (stretched vertically and reflected about the x-axis if c is negative).
- If $y = f(x)$ is replaced by $y = f(-x)$, the graph is reflected about the y-axis.

Shifting the graph of $y = f(x)$.

2.27

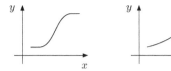

Graphs of increasing and strictly increasing functions.

2.28

Graphs of decreasing and strictly decreasing functions.

2.29

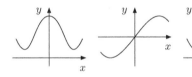

Graphs of even and odd functions, and of a function symmetric about $x = a$.

2.30

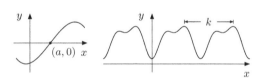

Graphs of a function symmetric about the point $(a, 0)$ and of a function periodic with period k.

2.31

$y = ax + b$ is a *nonvertical asymptote* for the curve $y = f(x)$ if

$$\lim_{x \to \infty} \big(f(x) - (ax + b)\big) = 0$$

or

$$\lim_{x \to -\infty} \big(f(x) - (ax + b)\big) = 0$$

Definition of a nonvertical asymptote.

2.32

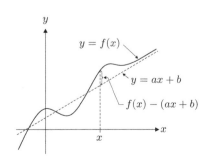

$y = ax + b$ is an asymptote for the curve $y = f(x)$.

How to find a nonvertical asymptote for the curve $y = f(x)$ as $x \to \infty$:

- Examine $\lim_{x \to \infty} \big(f(x)/x\big)$. If the limit does not exist, there is no asymptote as $x \to \infty$.

2.33
- If $\lim_{x \to \infty} \big(f(x)/x\big) = a$, examine the limit $\lim_{x \to \infty} \big(f(x) - ax\big)$. If this limit does not exist, the curve has no asymptote as $x \to \infty$.

- If $\lim_{x \to \infty} \big(f(x) - ax\big) = b$, then $y = ax + b$ is an *asymptote* for the curve $y = f(x)$ as $x \to \infty$.

Method for finding nonvertical asymptotes for a curve $y = f(x)$ as $x \to \infty$. Replacing $x \to \infty$ by $x \to -\infty$ gives a method for finding nonvertical asymptotes as $x \to -\infty$.

To find an approximate root of $f(x) = 0$, define x_n for $n = 1, 2, \ldots$, by

2.34
$$x_{n+1} = x_n - \frac{f(x_n)}{f'(x_n)}$$

If x_0 is close to an actual root x^*, the sequence $\{x_n\}$ will usually converge rapidly to that root.

Newton's approximation method. (A rule of thumb says that, to obtain an approximation that is correct to n decimal places, use Newton's method until it gives the same n decimal places twice in a row.)

2.35

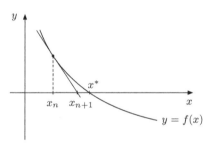

Illustration of Newton's approximation method. The tangent to the graph of f at $(x_n, f(x_n))$ intersects the x-axis at $x = x_{n+1}$.

2.36
Suppose in (2.34) that $f(x^*) = 0$, $f'(x^*) \neq 0$, and that $f''(x^*)$ exists and is continuous in a neighbourhood of x^*. Then there exists a $\delta > 0$ such that the sequence $\{x_n\}$ in (2.34) converges to x^* when $x_0 \in (x^* - \delta, x^* + \delta)$.

Sufficient conditions for convergence of Newton's method.

Suppose in (2.34) that f is twice differentiable with $f(x^*) = 0$ and $f'(x^*) \neq 0$. Suppose further that there exist a $K > 0$ and a $\delta > 0$ such that for all x in $(x^* - \delta, x^* + \delta)$,

2.37
$$\frac{|f(x)f''(x)|}{f'(x)^2} \leq K|x - x^*| < 1$$

Then if $x_0 \in (x^* - \delta, x^* + \delta)$, the sequence $\{x_n\}$ in (2.34) converges to x^* and

$$|x_n - x^*| \leq (\delta K)^{2^n}/K$$

A precise estimation of the accuracy of Newton's method.

2.38 $\quad y - f(x_1) = f'(x_1)(x - x_1)$

The equation for the *tangent* to $y = f(x)$ at $(x_1, f(x_1))$.

2.39 $\quad y - f(x_1) = -\dfrac{1}{f'(x_1)}(x - x_1)$

The equation for the *normal* to $y = f(x)$ at $(x_1, f(x_1))$.

2.40

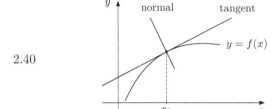

The tangent and the normal to $y = f(x)$ at $(x_1, f(x_1))$.

2.41

(i) $\quad a^r \cdot a^s = a^{r+s}$ $\qquad$ (ii) $\quad (a^r)^s = a^{rs}$

(iii) $\quad (ab)^r = a^r b^r$ $\qquad$ (iv) $\quad a^r / a^s = a^{r-s}$

(v) $\quad \left(\dfrac{a}{b}\right)^r = \dfrac{a^r}{b^r}$ $\qquad$ (vi) $\quad a^{-r} = \dfrac{1}{a^r}$

Rules for powers. (r and s are arbitrary real numbers, a and b are positive real numbers.)

2.42

- $e = \lim\limits_{n\to\infty} \left(1 + \dfrac{1}{n}\right)^n = 2.718281828459\ldots$

- $e^x = \lim\limits_{n\to\infty} \left(1 + \dfrac{x}{n}\right)^n$

- $\lim\limits_{n\to\infty} a_n = a \;\Rightarrow\; \lim\limits_{n\to\infty} \left(1 + \dfrac{a_n}{n}\right)^n = e^a$

Important definitions and results. See (8.23) for another formula for e^x.

2.43 $\quad e^{\ln x} = x$

Definition of the natural logarithm.

2.44

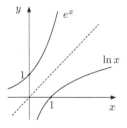

The graphs of $y = e^x$ and $y = \ln x$ are symmetric about the line $y = x$.

2.45

$\ln(xy) = \ln x + \ln y; \quad \ln\dfrac{x}{y} = \ln x - \ln y$

$\ln x^p = p \ln x; \quad \ln\dfrac{1}{x} = -\ln x$

Rules for the natural logarithm function. (x and y are positive.)

2.46 $\quad a^{\log_a x} = x \quad (a > 0,\ a \neq 1)$

Definition of the *logarithm* to the base a.

2.47
$$\log_a x = \frac{\ln x}{\ln a}; \quad \log_a b \cdot \log_b a = 1$$
$$\log_e x = \ln x; \quad \log_{10} x = \log_{10} e \cdot \ln x$$

Logarithms with different bases.

2.48
$$\log_a(xy) = \log_a x + \log_a y$$
$$\log_a \frac{x}{y} = \log_a x - \log_a y$$
$$\log_a x^p = p \log_a x, \quad \log_a \frac{1}{x} = -\log_a x$$

Rules for logarithms. (x and y are positive.)

2.49
$$1° = \frac{\pi}{180} \text{ rad}, \quad 1 \text{ rad} = \left(\frac{180}{\pi}\right)°$$

Relationship between degrees and radians (rad).

2.50

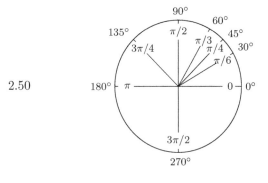

Relations between degrees and radians.

2.51

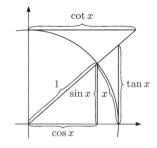

Definitions of the basic *trigonometric* functions. x is the length of the arc, and also the radian measure of the angle.

2.52

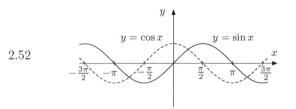

The graphs of $y = \sin x$ (—) and $y = \cos x$ (---). The functions sin and cos are periodic with period 2π:
$\sin(x + 2\pi) = \sin x$,
$\cos(x + 2\pi) = \cos x$.

2.53
$$\tan x = \frac{\sin x}{\cos x}, \quad \cot x = \frac{\cos x}{\sin x} = \frac{1}{\tan x}$$

Definition of the *tangent* and *cotangent* functions.

2.54

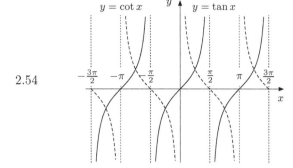

The graphs of $y = \tan x$ (—) and $y = \cot x$ (---). The functions tan and cot are periodic with period π:
$\tan(x + \pi) = \tan x$,
$\cot(x + \pi) = \cot x$.

2.55

x	0	$\frac{\pi}{6} = 30°$	$\frac{\pi}{4} = 45°$	$\frac{\pi}{3} = 60°$	$\frac{\pi}{2} = 90°$
$\sin x$	0	$\frac{1}{2}$	$\frac{1}{2}\sqrt{2}$	$\frac{1}{2}\sqrt{3}$	1
$\cos x$	1	$\frac{1}{2}\sqrt{3}$	$\frac{1}{2}\sqrt{2}$	$\frac{1}{2}$	0
$\tan x$	0	$\frac{1}{3}\sqrt{3}$	1	$\sqrt{3}$	*
$\cot x$	*	$\sqrt{3}$	1	$\frac{1}{3}\sqrt{3}$	0

* not defined

Special values of the trigonometric functions.

2.56

x	$\frac{3\pi}{4} = 135°$	$\pi = 180°$	$\frac{3\pi}{2} = 270°$	$2\pi = 360°$
$\sin x$	$\frac{1}{2}\sqrt{2}$	0	-1	0
$\cos x$	$-\frac{1}{2}\sqrt{2}$	-1	0	1
$\tan x$	-1	0	*	0
$\cot x$	-1	*	0	*

* not defined

2.57 $\displaystyle\lim_{x \to 0} \frac{\sin ax}{x} = a$

An important limit.

2.58 $\sin^2 x + \cos^2 x = 1$

Trigonometric formulas. (For series expansions of trigonometric functions, see Chapter 8.)

2.59 $\tan^2 x = \dfrac{1}{\cos^2 x} - 1, \qquad \cot^2 x = \dfrac{1}{\sin^2 x} - 1$

2.60
$\cos(x + y) = \cos x \cos y - \sin x \sin y$
$\cos(x - y) = \cos x \cos y + \sin x \sin y$
$\sin(x + y) = \sin x \cos y + \cos x \sin y$
$\sin(x - y) = \sin x \cos y - \cos x \sin y$

2.61
$$\tan(x + y) = \frac{\tan x + \tan y}{1 - \tan x \tan y}$$
$$\tan(x - y) = \frac{\tan x - \tan y}{1 + \tan x \tan y}$$

Trigonometric formulas.

2.62
$$\cos 2x = 2\cos^2 x - 1 = 1 - 2\sin^2 x$$
$$\sin 2x = 2\sin x \cos x$$

2.63 $\sin^2 \dfrac{x}{2} = \dfrac{1 - \cos x}{2}$, $\qquad \cos^2 \dfrac{x}{2} = \dfrac{1 + \cos x}{2}$

2.64
$$\cos x + \cos y = 2\cos \frac{x + y}{2} \cos \frac{x - y}{2}$$
$$\cos x - \cos y = -2\sin \frac{x + y}{2} \sin \frac{x - y}{2}$$

2.65
$$\sin x + \sin y = 2\sin \frac{x + y}{2} \cos \frac{x - y}{2}$$
$$\sin x - \sin y = 2\cos \frac{x + y}{2} \sin \frac{x - y}{2}$$

2.66
$$y = \arcsin x \Leftrightarrow x = \sin y, \ x \in [-1, 1], \ y \in \left[-\frac{\pi}{2}, \frac{\pi}{2}\right]$$
$$y = \arccos x \Leftrightarrow x = \cos y, \ x \in [-1, 1], \ y \in [0, \pi]$$
$$y = \arctan x \Leftrightarrow x = \tan y, \ x \in \mathbb{R}, \ y \in \left(-\frac{\pi}{2}, \frac{\pi}{2}\right)$$
$$y = \operatorname{arccot} x \Leftrightarrow x = \cot y, \ x \in \mathbb{R}, \ y \in (0, \pi)$$

Definitions of the inverse trigonometric functions.

2.67

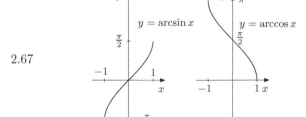

Graphs of the inverse trigonometric functions $y = \arcsin x$ and $y = \arccos x$.

2.68

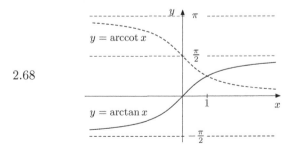

Graphs of the inverse trigonometric functions $y = \arctan x$ and $y = \operatorname{arccot} x$.

2.69

$$\arcsin x = \sin^{-1} x, \quad \arccos x = \cos^{-1} x$$
$$\arctan x = \tan^{-1} x, \quad \operatorname{arccot} x = \cot^{-1} x$$

Alternative notation for the inverse trigonometric functions.

2.70

$$\arcsin(-x) = -\arcsin x$$
$$\arccos(-x) = \pi - \arccos x$$
$$\arctan(-x) = \arctan x$$
$$\operatorname{arccot}(-x) = \pi - \operatorname{arccot} x$$
$$\arcsin x + \arccos x = \frac{\pi}{2}$$
$$\arctan x + \operatorname{arccot} x = \frac{\pi}{2}$$
$$\arctan \frac{1}{x} = \frac{\pi}{2} - \arctan x, \quad x > 0$$
$$\arctan \frac{1}{x} = -\frac{\pi}{2} - \arctan x, \quad x < 0$$

Properties of the inverse trigonometric functions.

2.71

$$\sinh x = \frac{e^x - e^{-x}}{2}, \qquad \cosh x = \frac{e^x + e^{-x}}{2}$$

Hyperbolic sine and cosine.

2.72

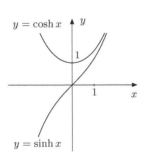

Graphs of the hyperbolic functions $y = \sinh x$ and $y = \cosh x$.

2.73

$$\cosh^2 x - \sinh^2 x = 1$$
$$\cosh(x + y) = \cosh x \cosh y + \sinh x \sinh y$$
$$\cosh 2x = \cosh^2 x + \sinh^2 x$$
$$\sinh(x + y) = \sinh x \cosh y + \cosh x \sinh y$$
$$\sinh 2x = 2 \sinh x \cosh x$$

Properties of hyperbolic functions.

2.74

$$y = \operatorname{arsinh} x \iff x = \sinh y$$
$$y = \operatorname{arcosh} x, \ x \geq 1 \iff x = \cosh y, \ y \geq 0$$
$$\operatorname{arsinh} x = \ln\left(x + \sqrt{x^2 + 1}\right)$$
$$\operatorname{arcosh} x = \ln\left(x + \sqrt{x^2 - 1}\right), \ x \geq 1$$

Definition of the inverse hyperbolic functions.

Complex numbers

2.75 $\quad z = a + ib, \quad \bar{z} = a - ib$

A *complex number* and its *conjugate*. $a, b \in \mathbb{R}$, and $i^2 = -1$. i is called the *imaginary unit*.

2.76 $\quad |z| = \sqrt{a^2 + b^2}, \quad \mathrm{Re}(z) = a, \quad \mathrm{Im}(z) = b$

$|z|$ is the *modulus* of $z = a + ib$. $\mathrm{Re}(z)$ and $\mathrm{Im}(z)$ are the *real* and *imaginary parts* of z.

2.77

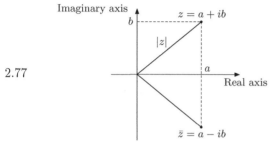

Geometric representation of a complex number and its conjugate.

2.78
- $(a + ib) + (c + id) = (a + c) + i(b + d)$
- $(a + ib) - (c + id) = (a - c) + i(b - d)$
- $(a + ib)(c + id) = (ac - bd) + i(ad + bc)$
- $\dfrac{a + ib}{c + id} = \dfrac{1}{c^2 + d^2}\big((ac + bd) + i(bc - ad)\big)$

Addition, subtraction, multiplication, and *division* of complex numbers.

2.79 $\quad |\bar{z}_1| = |z_1|, \quad z_1 \bar{z}_1 = |z_1|^2, \quad \overline{z_1 + z_2} = \bar{z}_1 + \bar{z}_2,$
$|z_1 z_2| = |z_1||z_2|, \quad |z_1 + z_2| \le |z_1| + |z_2|$

Basic rules. z_1 and z_2 are complex numbers.

2.80 $\quad z = a + ib = r(\cos\theta + i\sin\theta) = re^{i\theta}, \text{ where}$
$r = |z| = \sqrt{a^2 + b^2}, \quad \cos\theta = \dfrac{a}{r}, \quad \sin\theta = \dfrac{b}{r}$

The *trigonometric* or *polar* form of a complex number. The angle θ is called the *argument* of z. See (2.84) for $e^{i\theta}$.

2.81

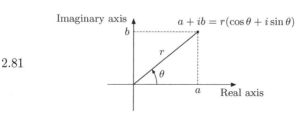

Geometric representation of the trigonometric form of a complex number.

If $z_k = r_k(\cos\theta_k + i\sin\theta_k)$, $k = 1, 2$, then

2.82
$$z_1 z_2 = r_1 r_2 \big(\cos(\theta_1 + \theta_2) + i\sin(\theta_1 + \theta_2)\big)$$
$$\frac{z_1}{z_2} = \frac{r_1}{r_2}\big(\cos(\theta_1 - \theta_2) + i\sin(\theta_1 - \theta_2)\big)$$

Multiplication and division on trigonometric form.

2.83 $(\cos\theta + i\sin\theta)^n = \cos n\theta + i\sin n\theta$

De Moivre's formula, $n = 0, 1, \ldots$.

If $z = x + iy$, then

2.84
$$e^z = e^{x+iy} = e^x \cdot e^{iy} = e^x(\cos y + i\sin y)$$
In particular,
$$e^{iy} = \cos y + i\sin y$$

The *complex exponential function*.

2.85 $e^{\pi i} = -1$

A striking relationship.

2.86
$$e^{\bar{z}} = \overline{e^z}, \quad e^{z+2\pi i} = e^z, \quad e^{z_1+z_2} = e^{z_1}e^{z_2},$$
$$e^{z_1-z_2} = e^{z_1}/e^{z_2}$$

Rules for the complex exponential function.

2.87 $\cos z = \dfrac{e^{iz} + e^{-iz}}{2}, \quad \sin z = \dfrac{e^{iz} - e^{-iz}}{2i}$

Euler's formulas.

If $a = r(\cos\theta + i\sin\theta) \neq 0$, then the equation
$$z^n = a$$

2.88 has exactly n roots, namely

$$z_k = \sqrt[n]{r}\Big(\cos\frac{\theta + 2k\pi}{n} + i\sin\frac{\theta + 2k\pi}{n}\Big)$$
for $k = 0, 1, \ldots, n-1$.

nth roots of a complex number, $n = 1, 2, \ldots$.

References

Most of these formulas can be found in any calculus text, e.g. Edwards and Penney (1998) or Sydsæter and Hammond (2005). For (2.3)–(2.12), see e.g. Turnbull (1952).

Limits

$$\lim_{t \to 1} \frac{t}{t-1} \Rightarrow \text{Limit does not exist}$$

$$\lim_{t \to 1} \frac{(\ln(t))^2}{(t^2-1)} = \lim_{t \to 1} \frac{\frac{d}{dt}((\ln(t))^2)}{\frac{d}{dt}(t^2-1)} = \lim_{t \to 1} \frac{2\ln(t)\left(\frac{1}{t}\right)}{2t} = \frac{0}{1} = 0$$

$$\lim_{\varepsilon \to 0} \frac{(5+\varepsilon)^2 - 5(2)}{\varepsilon} = \frac{\frac{d}{d\varepsilon}((5+\varepsilon)^2 - (5)^2)}{\frac{d}{d\varepsilon}\varepsilon} = \lim_{\varepsilon \to 0} \frac{2(5+\varepsilon)}{1} = 10$$

$$\lim_{\varepsilon \to 0} \frac{f(x+\varepsilon) - f(x)}{\varepsilon}$$

Chapter 3

Limits. Continuity. Differentiation (one variable)

3.1
$f(x)$ tends to A as a *limit* as x approaches a,
$$\lim_{x \to a} f(x) = A \quad \text{or} \quad f(x) \to A \text{ as } x \to a$$
if for every number $\varepsilon > 0$ there exists a number $\delta > 0$ such that
$$|f(x) - A| < \varepsilon \text{ if } x \in D_f \text{ and } 0 < |x - a| < \delta$$

The definition of a limit of a function of one variable. D_f is the domain of f.

3.2
If $\lim_{x \to a} f(x) = A$ and $\lim_{x \to a} g(x) = B$, then
- $\lim_{x \to a} \big(f(x) \pm g(x)\big) = A \pm B$
- $\lim_{x \to a} \big(f(x) \cdot g(x)\big) = A \cdot B$
- $\lim_{x \to a} \dfrac{f(x)}{g(x)} = \dfrac{A}{B} \quad$ (if $B \neq 0$)

Rules for limits.

3.3
f is *continuous* at $x = a$ if $\lim_{x \to a} f(x) = f(a)$, i.e. if $a \in D_f$ and for each number $\varepsilon > 0$ there is a number $\delta > 0$ such that
$$|f(x) - A| < \varepsilon \text{ if } x \in D_f \text{ and } |x - a| < \delta$$
f is *continuous on a set* $S \subset D_f$ if f is continuous at each point of S.

Definition of continuity.

3.4
If f and g are continuous at a, then:
- $f \pm g$ and $f \cdot g$ are continuous at a.
- f/g is continuous at a if $g(a) \neq 0$.

Properties of continuous functions.

3.5
If g is continuous at a, and f is continuous at $g(a)$, then $f(g(x))$ is continuous at a.

Continuity of composite functions.

3.6
Any function built from continuous functions by additions, subtractions, multiplications, divisions, and compositions, is continuous where defined.

A useful result.

3.7 f is *uniformly continuous* on a set S if for each $\varepsilon > 0$ there exists a $\delta > 0$ (depending on ε but NOT on x and y) such that
$$|f(x) - f(y)| < \varepsilon \text{ if } x,\, y \in S \text{ and } |x - y| < \delta$$

Definition of uniform continuity.

3.8 If f is continuous on a closed bounded interval I, then f is uniformly continuous on I.

Continuous functions on closed bounded intervals are uniformly continuous.

3.9 If f is continuous on an interval I containing a and b, and A lies between $f(a)$ and $f(b)$, then there is at least one ξ between a and b such that $A = f(\xi)$.

The *intermediate value theorem*.

3.10

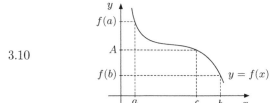

Illustration of the intermediate value theorem.

3.11 $f'(x) = \lim\limits_{h \to 0} \dfrac{f(x+h) - f(x)}{h}$

The definition of the *derivative*. If the limit exists, f is called *differentiable* at x.

3.12 Other notations for the derivative of $y = f(x)$ include
$$f'(x) = y' = \frac{dy}{dx} = \frac{df(x)}{dx} = Df(x)$$

Other notations for the derivative.

3.13 $y = f(x) \pm g(x) \;\Rightarrow\; y' = f'(x) \pm g'(x)$

General rules.

3.14 $y = f(x)g(x) \;\Rightarrow\; y' = f'(x)g(x) + f(x)g'(x)$

3.15 $y = \dfrac{f(x)}{g(x)} \;\Rightarrow\; y' = \dfrac{f'(x)g(x) - f(x)g'(x)}{\big(g(x)\big)^2}$

3.16 $y = f\big(g(x)\big) \;\Rightarrow\; y' = f'\big(g(x)\big) \cdot g'(x)$

The *chain rule*.

3.17 $y = f(x)^{g(x)} \;\Rightarrow$
$$y' = f(x)^{g(x)} \left(g'(x) \ln f(x) + g(x) \frac{f'(x)}{f(x)} \right)$$

A useful formula.

3.18 If $g = f^{-1}$ is the inverse of a one-to-one function f, and f is differentiable at x with $f'(x) \neq 0$, then g is differentiable at $f(x)$, and

$$g'(f(x)) = \frac{1}{f'(x)}$$

f^{-1} denotes the inverse function of f.

3.19

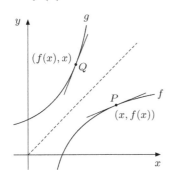

The graphs of f and $g = f^{-1}$ are symmetric with respect to the line $y = x$. If the slope of the tangent at P is $k = f'(x)$, then the slope $g'(f(x))$ of the tangent at Q equals $1/k$.

3.20 $y = c \Rightarrow y' = 0$ (c constant)

Special rules.

3.21 $y = x^a \Rightarrow y' = ax^{a-1}$ (a constant)

3.22 $y = \dfrac{1}{x} \Rightarrow y' = -\dfrac{1}{x^2}$

3.23 $y = \sqrt{x} \Rightarrow y' = \dfrac{1}{2\sqrt{x}}$

3.24 $y = e^x \Rightarrow y' = e^x$

3.25 $y = a^x \Rightarrow y' = a^x \ln a$ (a > 0)

3.26 $y = \ln x \Rightarrow y' = \dfrac{1}{x}$

3.27 $y = \log_a x \Rightarrow y' = \dfrac{1}{x} \log_a e$ (a > 0, a ≠ 1)

3.28 $y = \sin x \Rightarrow y' = \cos x$

3.29 $y = \cos x \Rightarrow y' = -\sin x$

3.30 $y = \tan x \Rightarrow y' = \dfrac{1}{\cos^2 x} = 1 + \tan^2 x$

3.31 $y = \cot x \Rightarrow y' = -\dfrac{1}{\sin^2 x} = -(1 + \cot^2 x)$

3.32 $\quad y = \sin^{-1} x = \arcsin x \;\Rightarrow\; y' = \dfrac{1}{\sqrt{1 - x^2}}$ $\qquad$ Special rules.

3.33 $\quad y = \cos^{-1} x = \arccos x \;\Rightarrow\; y' = -\dfrac{1}{\sqrt{1 - x^2}}$

3.34 $\quad y = \tan^{-1} x = \arctan x \;\Rightarrow\; y' = \dfrac{1}{1 + x^2}$

3.35 $\quad y = \cot^{-1} x = \operatorname{arccot} x \;\Rightarrow\; y' = -\dfrac{1}{1 + x^2}$

3.36 $\quad y = \sinh x \;\Rightarrow\; y' = \cosh x$

3.37 $\quad y = \cosh x \;\Rightarrow\; y' = \sinh x$

3.38 $\quad$ If f is continuous on $[a, b]$ and differentiable on (a, b), then there exists at least one point ξ in (a, b) such that

$$f'(\xi) = \frac{f(b) - f(a)}{b - a}$$

$\qquad$ The *mean value theorem*.

3.39

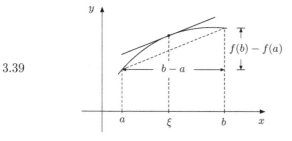

$\qquad$ Illustration of the mean value theorem.

3.40 $\quad$ If f and g are continuous on $[a, b]$ and differentiable on (a, b), then there exists at least one point ξ in (a, b) such that

$$[f(b) - f(a)]g'(\xi) = [g(b) - g(a)]f'(\xi)$$

$\qquad$ *Cauchy's generalized mean value theorem.*

3.41 $\quad$ Suppose f and g are differentiable on an interval (α, β) around a, except possibly at a, and suppose that $f(x)$ and $g(x)$ both tend to 0 as x tends to a. If $g'(x) \neq 0$ for all $x \neq a$ in (α, β) and $\lim_{x \to a} f'(x)/g'(x) = L$ (L finite, $L = \infty$ or $L = -\infty$), then

$$\lim_{x \to a} \frac{f(x)}{g(x)} = \lim_{x \to a} \frac{f'(x)}{g'(x)} = L$$

$\qquad$ *L'Hôpital's rule.* The same rule applies for $x \to a^+$, $x \to a^-$, $x \to \infty$, or $x \to -\infty$, and also if $f(x) \to \pm\infty$ and $g(x) \to \pm\infty$.

If $y = f(x)$ and dx is any number,

3.42 $\quad dy = f'(x)\,dx$

is the *differential* of y.

Definition of the differential.

3.43

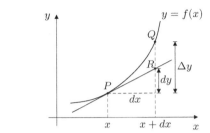

Geometric illustration of the differential.

3.44 $\quad \Delta y = f(x + dx) - f(x) \approx f'(x)\,dx$

when $|dx|$ is small.

A useful approximation, made more precise in (3.45).

3.45 $\quad f(x + dx) - f(x) = f'(x)\,dx + \varepsilon\,dx$

where $\varepsilon \to 0$ as $dx \to 0$

Property of a differentiable function. (If dx is very small, then ε is very small, and $\varepsilon\,dx$ is "very, very small".)

3.46
$$d(af + bg) = a\,df + b\,dg \quad (a \text{ and } b \text{ are constants})$$
$$d(fg) = g\,df + f\,dg$$
$$d(f/g) = (g\,df - f\,dg)/g^2$$
$$df(u) = f'(u)\,du$$

Rules for differentials. f and g are differentiable, and u is any differentiable function.

References

All formulas are standard and are found in almost any calculus text, e.g. Edwards and Penney (1998), or Sydsæter and Hammond (2005). For uniform continuity, see Rudin (1982).

$y = x_1 (x_2)^3 + \ln(x_2) x_3$

$dy = \dfrac{\partial f}{\partial x_1}(x)\,dx_1 + \dfrac{\partial f}{\partial x_2}(x)\,dx_2 + \dfrac{\partial f}{\partial x_3}(x)\,dx_3$

$dy = (x_2)^3\,dx_1 + \left(3x_1(x_2)^2 + \dfrac{x_3}{x_2}\right)dx_2 + \ln(v_2)\,dx_3$

Ex

$$f(x) = \left(\ln\left(\frac{1}{x^2}\right) - x\right)^3 + \frac{\exp(x)}{x^2}$$

$$f'(x) = 3\left(\ln\left(\frac{1}{x^2}\right) - x\right)^2\left(\left(1/\frac{1}{x^2}\right)(-2x^{-3}) - 1\right) + \exp(x) \cdot -2x^{-3} + \frac{\exp(x)}{x_2}$$

$$= 3\left(\ln\left(\frac{1}{x^2}\right) - x\right)^2\left(\frac{-2}{x} - 1\right) - \frac{2\exp(x)}{x^3} + \frac{\exp(x)}{x^2}$$

$$= -3\left(\ln\frac{1}{x^2} - x\right)^2\left(\frac{2}{x} + 1\right) + \frac{\exp(x)}{x^2}\left(1 - \frac{2}{x}\right)$$

Ex

$$f(x) = \exp\left(\ln\left(\frac{1}{x^2}\right) - 5\right), \text{ find } f''(x)$$

$$f'(x) = \exp\left(\ln\left(\frac{1}{x^2} - 5\right)\right)\left(\frac{1}{\frac{1}{x^2}}\right)\left(-2x^{-3}\right)$$

$$= -2\exp\left(\ln\left(\frac{1}{x^2}\right) - 5\right)\frac{1}{x}$$

$$f''(x) = -2\left(\exp\left(\ln\left(\frac{1}{x^2}\right) - 5\right)\frac{-1}{x^2} + \left(-2\exp\left(\ln\left(\frac{1}{x^2}\right) - 5\right)\frac{1}{x}\right)\frac{1}{x}\right)$$

$$= -2\exp\left(\ln\left(\frac{1}{x^2}\right) - 5\right)\left(-\frac{1}{x^2} - 2\frac{1}{x}\frac{1}{x}\right)$$

$$= -2\exp\left(\ln\left(\frac{1}{x^2}\right) - 5\right)\left(-\frac{1}{x^2} - \frac{2}{x^2}\right)$$

$$= 2\exp\left(\ln(1/x^2) - 5\right)\left(3/x^2\right)$$

$$= \frac{6}{x^2}\exp\left(\ln\left(\frac{1}{x^2}\right) - 5\right)$$

Total Derivative

Chapter 4

Partial derivatives

4.1 If $z = f(x_1, \ldots, x_n) = f(\mathbf{x})$, then
$$\frac{\partial z}{\partial x_i} = \frac{\partial f}{\partial x_i} = f_i'(\mathbf{x}) = D_{x_i} f = D_i f$$
all denote the derivative of $f(x_1, \ldots, x_n)$ with respect to x_i when all the other variables are held constant.

Definition of the *partial derivative*. (Other notations are also used.)

4.2

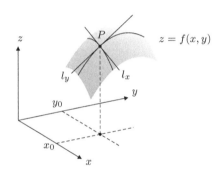

Geometric interpretation of the partial derivatives of a function of two variables, $z = f(x, y)$: $f_1'(x_0, y_0)$ is the slope of the tangent line l_x and $f_2'(x_0, y_0)$ is the slope of the tangent line l_y.

4.3
$$\frac{\partial^2 z}{\partial x_j \partial x_i} = f_{ij}''(x_1, \ldots, x_n) = \frac{\partial}{\partial x_j} f_i'(x_1, \ldots, x_n)$$

Second-order partial derivatives of $z = f(x_1, \ldots, x_n)$.

4.4
$$\frac{\partial^2 f}{\partial x_j \partial x_i} = \frac{\partial^2 f}{\partial x_i \partial x_j}, \quad i, j = 1, 2, \ldots, n$$

Young's theorem, valid if the two partials are continuous.

4.5 $f(x_1, \ldots, x_n)$ is said to be *of class* C^k, or simply C^k, in the set $S \subset \mathbb{R}^n$ if all partial derivatives of f of order $\leq k$ are continuous in S.

Definition of a C^k function. (For the definition of continuity, see (12.14).)

4.6
$$z = F(x, y), \ x = f(t), \ y = g(t) \Rightarrow$$
$$\frac{dz}{dt} = F_1'(x, y)\frac{dx}{dt} + F_2'(x, y)\frac{dy}{dt}$$

A *chain rule*.

4.7

If $z = F(x_1, \ldots, x_n)$ and $x_i = f_i(t_1, \ldots, t_m)$, $i = 1, \ldots, n$, then for all $j = 1, \ldots, m$

$$\frac{\partial z}{\partial t_j} = \sum_{i=1}^{n} \frac{\partial F(x_1, \ldots, x_n)}{\partial x_i} \frac{\partial x_i}{\partial t_j}$$

The chain rule. (General case.)

4.8

If $z = f(x_1, \ldots, x_n)$ and $dx_1, \ldots, dx_n$ are arbitrary numbers,

$$dz = \sum_{i=1}^{n} f_i'(x_1, \ldots, x_n) \, dx_i$$

is the *differential* of z.

Definition of the differential.

4.9

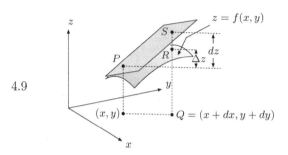

Geometric illustration of the definition of the differential for functions of two variables. It also illustrates the approximation $\Delta z \approx dz$ in (4.10).

4.10

$\Delta z \approx dz$ when $|dx_1|, \ldots, |dx_n|$ are all small, where

$$\Delta z = f(x_1 + dx_1, \ldots, x_n + dx_n) - f(x_1, \ldots, x_n)$$

A useful approximation, made more precise for differentiable functions in (4.11).

4.11

f is *differentiable* at $\mathbf{x}$ if $f_i'(\mathbf{x})$ all exist and there exist functions $\varepsilon_i = \varepsilon_i(dx_1, \ldots, dx_n)$, $i = 1, \ldots, n$, that all approach zero as dx_i all approach zero, and such that

$$\Delta z - dz = \varepsilon_1 \, dx_1 + \cdots + \varepsilon_n \, dx_n$$

Definition of differentiability.

4.12

If f is a C^1 function, i.e. it has continuous first order partials, then f is differentiable.

An important fact.

4.13

$$d(af + bg) = a \, df + b \, dg \quad (a \text{ and } b \text{ constants})$$
$$d(fg) = g \, df + f \, dg$$
$$d(f/g) = (g \, df - f \, dg)/g^2$$
$$dF(u) = F'(u) \, du$$

Rules for differentials. f and g are differentiable functions of x_1, $\ldots$, x_n, F is a differentiable function of one variable, and u is any differentiable function of $x_1, \ldots, x_n$.

4.14 $\quad F(x,y) = c \;\Rightarrow\; \dfrac{dy}{dx} = -\dfrac{F_1'(x,y)}{F_2'(x,y)}$

Formula for the slope of a level curve for $z = F(x,y)$. For precise assumptions, see (4.17).

4.15

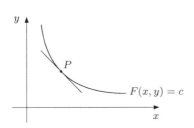

The slope of the tangent at P is
$$\frac{dy}{dx} = -\frac{F_1'(x,y)}{F_2'(x,y)}.$$

4.16

If $y = f(x)$ is a C^2 function satisfying $F(x,y) = c$, then

$$f''(x) = -\frac{F_{11}''(F_2')^2 - 2F_{12}''F_1'F_2' + F_{22}''(F_1')^2}{(F_2')^3}$$

$$= \frac{1}{(F_2')^3}\begin{vmatrix} 0 & F_1' & F_2' \\ F_1' & F_{11}'' & F_{12}'' \\ F_2' & F_{12}'' & F_{22}'' \end{vmatrix}$$

A useful result. All partials are evaluated at (x,y).

4.17

If $F(x,y)$ is C^k in a set A, (x_0, y_0) is an interior point of A, $F(x_0, y_0) = c$, and $F_2'(x_0, y_0) \neq 0$, then the equation $F(x,y) = c$ defines y as a C^k function of x, $y = \varphi(x)$, in some neighborhood of (x_0, y_0), and the derivative of y is

$$\frac{dy}{dx} = -\frac{F_1'(x,y)}{F_2'(x,y)}$$

The *implicit function theorem*. (For a more general result, see (6.3).)

4.18

If $F(x_1, x_2, \ldots, x_n, z) = c$ (c constant), then

$$\frac{\partial z}{\partial x_i} = -\frac{\partial F/\partial x_i}{\partial F/\partial z}, \quad i = 1, 2, \ldots, n \quad \left(\frac{\partial F}{\partial z} \neq 0\right)$$

A generalization of (4.14).

Homogeneous and homothetic functions

4.19

$f(\mathbf{x}) = f(x_1, x_2, \ldots, x_n)$ is *homogeneous of degree k* in $D \subset \mathbb{R}^n$ if
$$f(tx_1, tx_2, \ldots, tx_n) = t^k f(x_1, x_2, \ldots, x_n)$$
for all $t > 0$ and all $\mathbf{x} = (x_1, x_2, \ldots, x_n)$ in D.

The definition of a homogeneous function. D is a *cone* in the sense that $tx \in D$ whenever $x \in D$ and $t > 0$.

EX, $u_1(x) = x_1 x_2^2$ EX, $u_2(x_1,x_2) = (x_1)^{1/3}(x_2)^{2/3}$ EX, $u_3(x_1,x_2) = \frac{1}{3}\ln(x_1) + \frac{2}{3}\ln(x_2)$

HOD-3   HOD-1   NUT HOD.

4.20

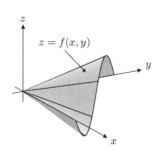

Geometric illustration of a function homogeneous of degree 1. (Only a portion of the graph is shown.)

4.21

$f(\mathbf{x}) = f(x_1, \dots, x_n)$ is homogeneous of degree k in the open cone D if and only if

$$\sum_{i=1}^{n} x_i f_i'(\mathbf{x}) = k f(\mathbf{x}) \text{ for all } \mathbf{x} \text{ in } D$$

Euler's theorem, valid for C^1 functions.

4.22

If $f(\mathbf{x}) = f(x_1, \dots, x_n)$ is homogeneous of degree k in the open cone D, then

- $\partial f / \partial x_i$ is homogeneous of degree $k - 1$ in D

- $\sum_{i=1}^{n} \sum_{j=1}^{n} x_i x_j f_{ij}''(\mathbf{x}) = k(k - 1) f(\mathbf{x})$

Properties of homogeneous functions.

4.23

$f(\mathbf{x}) = f(x_1, \dots, x_n)$ is *homothetic* in the cone D if for all $\mathbf{x}, \mathbf{y} \in D$ and all $t > 0$,

$$f(\mathbf{x}) = f(\mathbf{y}) \;\Rightarrow\; f(t\mathbf{x}) = f(t\mathbf{y})$$

Definition of homothetic function.

4.24

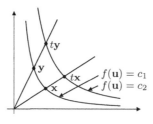

Geometric illustration of a homothetic function. With $f(\mathbf{u})$ homothetic, if $\mathbf{x}$ and $\mathbf{y}$ are on the same level curve, then so are $t\mathbf{x}$ and $t\mathbf{y}$ (when $t > 0$).

4.25

Let $f(\mathbf{x})$ be a continuous, homothetic function defined in a connected cone D. Assume that f is strictly increasing along each ray in D, i.e. for each $\mathbf{x}_0 \neq \mathbf{0}$ in D, $f(t\mathbf{x}_0)$ is a strictly increasing function of t. Then there exist a homogeneous function g and a strictly increasing function F such that

$$f(\mathbf{x}) = F(g(\mathbf{x})) \text{ for all } \mathbf{x} \text{ in } D$$

A property of continuous, homothetic functions (which is sometimes taken as the definition of homotheticity). One can assume that g is homogeneous of degree 1.

Gradients, directional derivatives, and tangent planes

4.26 $\nabla f(\mathbf{x}) = \left(\dfrac{\partial f(\mathbf{x})}{\partial x_1}, \ldots, \dfrac{\partial f(\mathbf{x})}{\partial x_n} \right)$

The *gradient* of f at $\mathbf{x} = (x_1, \ldots, x_n)$.

4.27 $f'_{\mathbf{a}}(\mathbf{x}) = \lim\limits_{h \to 0} \dfrac{f(\mathbf{x} + h\mathbf{a}) - f(\mathbf{x})}{h}, \quad \|\mathbf{a}\| = 1$

The *directional derivative* of f at $\mathbf{x}$ in the direction $\mathbf{a}$.

4.28 $f'_{\mathbf{a}}(\mathbf{x}) = \sum\limits_{i=1}^{n} f'_i(\mathbf{x}) a_i = \nabla f(\mathbf{x}) \cdot \mathbf{a}$

The relationship between the directional derivative and the gradient.

4.29
- $\nabla f(\mathbf{x})$ is orthogonal to the level surface $f(\mathbf{x}) = C$.
- $\nabla f(\mathbf{x})$ points in the direction of maximal increase of f.
- $\|\nabla f(\mathbf{x})\|$ measures the rate of change of f in the direction of $\nabla f(\mathbf{x})$.

Properties of the gradient.

4.30

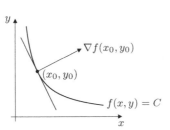

The gradient $\nabla f(x_0, y_0)$ of $f(x, y)$ at (x_0, y_0).

4.31 The *tangent plane* to the graph of $z = f(x, y)$ at the point $P = (x_0, y_0, z_0)$, with $z_0 = f(x_0, y_0)$, has the equation

$$z - z_0 = f'_1(x_0, y_0)(x - x_0) + f'_2(x_0, y_0)(y - y_0)$$

Definition of the tangent plane.

4.32

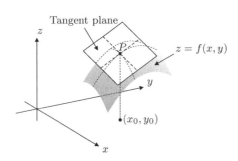

The graph of a function and its tangent plane.

The *tangent hyperplane* to the level surface

$$F(\mathbf{x}) = F(x_1, \ldots, x_n) = C$$

at the point $\mathbf{x}^0 = (x_1^0, \ldots, x_n^0)$ has the equation

$$\nabla F(\mathbf{x}^0) \cdot (\mathbf{x} - \mathbf{x}^0) = 0$$

4.33

> Definition of the tangent hyperplane. The vector $\nabla F(\mathbf{x}^0)$ is a *normal* to the hyperplane.

Let f be defined on a convex set $S \subset \mathbb{R}^n$, and let $\mathbf{x}^0$ be an interior point in S.

- If f is concave, there is at least one vector $\mathbf{p}$ in $\mathbb{R}^n$ such that

4.34

$$f(\mathbf{x}) - f(\mathbf{x}^0) \leq \mathbf{p} \cdot (\mathbf{x} - \mathbf{x}^0) \quad \text{for all } \mathbf{x} \text{ in } S$$

- If f is convex, there is at least one vector $\mathbf{p}$ in $\mathbb{R}^n$ such that

$$f(\mathbf{x}) - f(\mathbf{x}^0) \geq \mathbf{p} \cdot (\mathbf{x} - \mathbf{x}^0) \quad \text{for all } \mathbf{x} \text{ in } S$$

> A vector $\mathbf{p}$ that satisfies the first inequality is called a *supergradient* for f at $\mathbf{x}^0$. A vector satisfying the second inequality is called a *subgradient* for f at $\mathbf{x}^0$.

If f is defined on a set $S \subset \mathbb{R}^n$ and $\mathbf{x}^0$ is an interior point in S at which f is differentiable and $\mathbf{p}$ is a vector that satisfies either inequality in (4.34), then $\mathbf{p} = \nabla f(\mathbf{x}^0)$.

4.35

> A useful result.

Differentiability for mappings from $\mathbb{R}^n$ to $\mathbb{R}^m$

A transformation $\mathbf{f} = (f_1, \ldots, f_m)$ from a subset A of $\mathbb{R}^n$ into $\mathbb{R}^m$ is *differentiable* at an interior point $\mathbf{x}$ of A if (and only if) each component function $f_i : A \to \mathbb{R}$, $i = 1, \ldots m$, is differentiable at $\mathbf{x}$. Moreover, we define the derivative of $\mathbf{f}$ at $\mathbf{x}$ by

4.36

$$\mathbf{f}'(\mathbf{x}) = \begin{pmatrix} \dfrac{\partial f_1}{\partial x_1}(\mathbf{x}) & \dfrac{\partial f_1}{\partial x_2}(\mathbf{x}) & \cdots & \dfrac{\partial f_1}{\partial x_n}(\mathbf{x}) \\ \vdots & \vdots & & \vdots \\ \dfrac{\partial f_m}{\partial x_1}(\mathbf{x}) & \dfrac{\partial f_m}{\partial x_2}(\mathbf{x}) & \cdots & \dfrac{\partial f_m}{\partial x_n}(\mathbf{x}) \end{pmatrix}$$

the $m \times n$ matrix whose ith row is $f_i'(\mathbf{x}) = \nabla f_i(\mathbf{x})$.

> Generalizes (4.11).

If a transformation $\mathbf{f}$ from $A \subset \mathbb{R}^n$ into $\mathbb{R}^m$ is differentiable at an interior point $\mathbf{a}$ of A, then $\mathbf{f}$ is continuous at $\mathbf{a}$.

4.37

> Differentiability implies continuity.

4.38	A transformation $\mathbf{f} = (f_1, \ldots, f_m)$ from (a subset of) $\mathbb{R}^n$ into $\mathbb{R}^m$ is said to be of *class* C^k if each of its component functions $f_1, \ldots, f_m$ is C^k.	An important definition. (See (4.5).)
4.39	If $\mathbf{f}$ is a C^1 transformation from an open set $A \subset \mathbb{R}^n$ into $\mathbb{R}^m$, then $\mathbf{f}$ is differentiable at every point $\mathbf{x}$ in A.	C^1 transformations are differentiable.
4.40	Suppose $\mathbf{f} : A \to \mathbb{R}^m$ and $\mathbf{g} : B \to \mathbb{R}^p$ are defined on $A \subset \mathbb{R}^n$ and $B \subset \mathbb{R}^m$, with $\mathbf{f}(A) \subset B$, and suppose that $\mathbf{f}$ and $\mathbf{g}$ are differentiable at $\mathbf{x}$ and $\mathbf{f}(\mathbf{x})$, respectively. Then the composite transformation $\mathbf{g} \circ \mathbf{f} : A \to \mathbb{R}^p$ defined by $(\mathbf{g} \circ \mathbf{f})(\mathbf{x}) = \mathbf{g}(\mathbf{f}(\mathbf{x}))$ is differentiable at $\mathbf{x}$, and $$(\mathbf{g} \circ \mathbf{f})'(\mathbf{x}) = \mathbf{g}'(\mathbf{f}(\mathbf{x}))\,\mathbf{f}'(\mathbf{x})$$	The chain rule.

References

Most of the formulas are standard and can be found in almost any calculus text, e.g. Edwards and Penney (1998), or Sydsæter and Hammond (2005). For supergradients and differentiability, see e.g. Sydsæter et al. (2005). For properties of homothetic functions, see Simon and Blume (1994), Shephard (1970), and Førsund (1975).

Implicit Function Theorem Example

$x^2 y + x y^3 = 42 \quad (x, y) = (3, 2) \qquad 3^2(2) + 3(2)^3 = 42$
$$9(2) + 3(8) = 42$$
$$42 = 42$$

$\dfrac{\partial F}{\partial x}(x, y) = 2xy + y^3$
$\dfrac{\partial F}{\partial y}(x, y) = x^2 + 3y^2 x$ well defined, continuous vector (2nd condition)

3rd condition: $\dfrac{\partial F}{\partial y}(3, 2) = 3^2 + 3(3)(2)^2 = 45 \neq 0$

$\dfrac{dy}{dx} = -\dfrac{2xy + y^3}{x^2 + 3xy^2} = -\dfrac{2(3)(2) + 2^3}{3^2 + 3(3)(2)^2} = -\dfrac{4}{9}$

$\dfrac{dy}{dx} = -\dfrac{4}{9}$

Chapter 5

Elasticities. Elasticities of substitution

5.1 $\quad \mathrm{El}_x \, f(x) = \dfrac{x}{f(x)} f'(x) = \dfrac{x}{y} \dfrac{dy}{dx} = \dfrac{d(\ln y)}{d(\ln x)}$

$\mathrm{El}_x \, f(x)$, the *elasticity* of $y = f(x)$ w.r.t. x, is approximately the per-centage change in $f(x)$ corresponding to a one per cent increase in x.

5.2

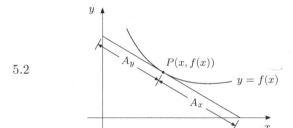

Illustration of Marshall's rule.

5.3 *Marshall's rule:* To find the elasticity of $y = f(x)$ w.r.t. x at the point P in the figure, first draw the tangent to the curve at P. Measure the distance A_y from P to the point where the tangent intersects the y-axis, and the distance A_x from P to where the tangent intersects the x-axis. Then $\mathrm{El}_x \, f(x) = \pm A_y/A_x$.

Marshall's rule. The distances are measured positive. Choose the plus sign if the curve is increasing at P, the minus sign in the opposite case.

5.4
- If $|\,\mathrm{El}_x \, f(x)\,| > 1$, then f is elastic at x.
- If $|\,\mathrm{El}_x \, f(x)\,| = 1$, then f is unitary elastic at x.
- If $|\,\mathrm{El}_x \, f(x)\,| < 1$, then f is inelastic at x.
- If $|\,\mathrm{El}_x \, f(x)\,| = 0$, then f is completely in-elastic at x.

Terminology used by many economists.

5.5 $\quad \mathrm{El}_x(f(x)g(x)) = \mathrm{El}_x \, f(x) + \mathrm{El}_x \, g(x)$

General rules for calculating elasticities.

5.6 $\quad \mathrm{El}_x \left(\dfrac{f(x)}{g(x)} \right) = \mathrm{El}_x \, f(x) - \mathrm{El}_x \, g(x)$

5.7 $\quad \mathrm{El}_x(f(x) \pm g(x)) = \dfrac{f(x)\,\mathrm{El}_x f(x) \pm g(x)\,\mathrm{El}_x g(x)}{f(x) \pm g(x)}$ | General rules for calculating elasticities.

5.8 $\quad \mathrm{El}_x f(g(x)) = \mathrm{El}_u f(u)\,\mathrm{El}_x u, \quad u = g(x)$ |

5.9 If $y = f(x)$ has an inverse function $x = g(y) = f^{-1}(y)$, then, with $y_0 = f(x_0)$,

$$\mathrm{El}_y x = \frac{y}{x}\frac{dx}{dy}, \quad \text{i.e.} \quad \mathrm{El}_y(g(y_0)) = \frac{1}{\mathrm{El}_x f(x_0)}$$

| The elasticity of the inverse function.

5.10 $\quad \mathrm{El}_x A = 0, \quad \mathrm{El}_x x^a = a, \quad \mathrm{El}_x e^x = x.$
(A and a are constants, $A \neq 0$.)

| Special rules for elasticities.

5.11 $\quad \mathrm{El}_x \sin x = x \cot x, \quad \mathrm{El}_x \cos x = -x \tan x$ |

5.12 $\quad \mathrm{El}_x \tan x = \dfrac{x}{\sin x \cos x}, \quad \mathrm{El}_x \cot x = \dfrac{-x}{\sin x \cos x}$ |

5.13 $\quad \mathrm{El}_x \ln x = \dfrac{1}{\ln x}, \quad \mathrm{El}_x \log_a x = \dfrac{1}{\ln x}$ |

5.14 $\quad \mathrm{El}_i f(\mathbf{x}) = \mathrm{El}_{x_i} f(\mathbf{x}) = \dfrac{x_i}{f(\mathbf{x})}\dfrac{\partial f(\mathbf{x})}{\partial x_i}$ | The *partial elasticity* of $f(\mathbf{x}) = f(x_1,\ldots,x_n)$ w.r.t. x_i, $i = 1,\ldots,n$.

5.15 If $z = F(x_1,\ldots,x_n)$ and $x_i = f_i(t_1,\ldots,t_m)$ for $i = 1,\ldots,n$, then for all $j = 1,\ldots,m$,

$$\mathrm{El}_{t_j} z = \sum_{i=1}^{n} \mathrm{El}_i F(x_1,\ldots,x_n)\,\mathrm{El}_{t_j} x_i$$

| The *chain rule for elasticities.*

5.16 The *directional elasticity* of f at $\mathbf{x}$, in the direction of $\mathbf{x}/\|\mathbf{x}\|$, is

$$\mathrm{El}_{\mathbf{a}} f(\mathbf{x}) = \frac{\|\mathbf{x}\|}{f(\mathbf{x})} f_{\mathbf{a}}'(\mathbf{x}) = \frac{1}{f(\mathbf{x})}\nabla f(\mathbf{x}) \cdot \mathbf{x}$$

| $\mathrm{El}_{\mathbf{a}} f(\mathbf{x})$ is approximately the percentage change in $f(\mathbf{x})$ corresponding to a one per cent increase in each component of $\mathbf{x}$. (See (4.27)–(4.28) for $f_{\mathbf{a}}'(\mathbf{x})$.)

5.17 $\quad \mathrm{El}_{\mathbf{a}} f(\mathbf{x}) = \displaystyle\sum_{i=1}^{n} \mathrm{El}_i f(\mathbf{x}), \quad \mathbf{a} = \dfrac{\mathbf{x}}{\|\mathbf{x}\|}$ | A useful fact (the *passus equation*).

5.18 The *marginal rate of substitution* (MRS) of y for x is

$$R_{yx} = \frac{f_1'(x,y)}{f_2'(x,y)}, \quad f(x,y) = c$$

| R_{yx} is approximately how much one must add of y per unit of x removed to stay on the same level curve for f.

- When f is a utility function, and x and y are goods, R_{yx} is called the *marginal rate of substitution* (abbreviated MRS).

5.19
- When f is a production function and x and y are inputs, R_{yx} is called the *marginal rate of technical substitution* (abbreviated MRTS).
- When $f(x, y) = 0$ is a production function in implicit form (for given factor inputs), and x and y are two products, R_{yx} is called the *marginal rate of product transformation* (abbreviated MRPT).

Different special cases of (5.18). See Chapters 25 and 26.

The *elasticity of substitution* between y and x is

5.20
$$\sigma_{yx} = \text{El}_{R_{yx}}\left(\frac{y}{x}\right) = -\frac{\partial \ln\left(\frac{y}{x}\right)}{\partial \ln\left(\frac{f_2'}{f_1'}\right)}, \quad f(x, y) = c$$

σ_{yx} is, approximately, the percentage change in the factor ratio y/x corresponding to a one percent change in the marginal rate of substitution, assuming that f is constant.

5.21
$$\sigma_{yx} = \frac{\dfrac{1}{xf_1'} + \dfrac{1}{yf_2'}}{-\dfrac{f_{11}''}{(f_1')^2} + 2\dfrac{f_{12}''}{f_1'f_2'} - \dfrac{f_{22}''}{(f_2')^2}}, \quad f(x, y) = c$$

An alternative formula for the elasticity of substitution. Note that $\sigma_{yx} = \sigma_{xy}$.

If $f(x, y)$ is homogeneous of degree 1, then

5.22
$$\sigma_{yx} = \frac{f_1' f_2'}{f f_{12}''}$$

A special case.

5.23
$$h_{ji}(\mathbf{x}) = \frac{\partial f(\mathbf{x})}{\partial x_i} \Big/ \frac{\partial f(\mathbf{x})}{\partial x_j}, \quad i, j = 1, 2, \ldots, n$$

The marginal rate of substitution of factor j for factor i.

5.24
If f is a strictly increasing transformation of a homogeneous function, as in (4.25), then the marginal rates of substitution in (5.23) are homogeneous of degree 0.

A useful result.

5.25
$$\sigma_{ij} = -\frac{\partial \ln\left(\frac{x_i}{x_j}\right)}{\partial \ln\left(\frac{f_i'}{f_j'}\right)}, \quad f(x_1, \ldots, x_n) = c, \ i \neq j$$

The elasticity of substitution in the n-variable case.

5.26 $\quad \sigma_{ij} = \dfrac{\dfrac{1}{x_i f_i'} + \dfrac{1}{x_j f_j'}}{-\dfrac{f_{ii}''}{(f_i')^2} + \dfrac{2f_{ij}''}{f_i' f_j'} - \dfrac{f_{jj}''}{(f_j')^2}}, \quad i \neq j$

The *elasticity of substitution*, $f(x_1, \ldots, x_n) = c$.

References

These formulas will usually not be found in calculus texts. For (5.5)–(5.24), see e.g. Sydsæter and Hammond (2005). For (5.25)–(5.26), see Blackorby and Russell (1989) and Fuss and McFadden (1978). For elasticities of substitution in production theory, see Chapter 25.

Chapter 6

Systems of equations

6.1
$$f_1(x_1, x_2, \ldots, x_n, y_1, y_2, \ldots, y_m) = 0$$
$$f_2(x_1, x_2, \ldots, x_n, y_1, y_2, \ldots, y_m) = 0$$
$$\cdots\cdots\cdots\cdots\cdots\cdots\cdots\cdots\cdots\cdots\cdots$$
$$f_m(x_1, x_2, \ldots, x_n, y_1, y_2, \ldots, y_m) = 0$$

A general system of equations with n *exogenous variables*, $x_1, \ldots, x_n$, and m *endogenous variables*, $y_1, \ldots, y_m$.

6.2
$$\frac{\partial \mathbf{f}(\mathbf{x}, \mathbf{y})}{\partial \mathbf{y}} = \begin{pmatrix} \dfrac{\partial f_1}{\partial y_1} & \cdots & \dfrac{\partial f_1}{\partial y_m} \\ \vdots & \ddots & \vdots \\ \dfrac{\partial f_m}{\partial y_1} & \cdots & \dfrac{\partial f_m}{\partial y_m} \end{pmatrix}$$

The *Jacobian matrix* of $f_1, \ldots, f_m$ with respect to $y_1, \ldots, y_m$.

6.3
Suppose $f_1, \ldots, f_m$ are C^k functions in a set A in $\mathbb{R}^{n+m}$, let $(\mathbf{x}^0, \mathbf{y}^0) = (x_1^0, \ldots, x_n^0, y_1^0, \ldots, y_m^0)$ be a solution to (6.1) in the interior of A. Suppose also that the determinant of the Jacobian matrix $\partial \mathbf{f}(\mathbf{x}, \mathbf{y})/\partial \mathbf{y}$ in (6.2) is different from 0 at $(\mathbf{x}^0, \mathbf{y}^0)$. Then (6.1) defines $y_1, \ldots, y_m$ as C^k functions of $x_1, \ldots, x_n$ in some neighborhood of $(\mathbf{x}^0, \mathbf{y}^0)$, and the Jacobian matrix of these functions with respect to $\mathbf{x}$ is

$$\frac{\partial \mathbf{y}}{\partial \mathbf{x}} = \left(\frac{\partial \mathbf{f}(\mathbf{x}, \mathbf{y})}{\partial \mathbf{y}} \right)^{-1} \frac{\partial \mathbf{f}(\mathbf{x}, \mathbf{y})}{\partial \mathbf{x}}$$

The *general implicit function theorem*. (It gives sufficient conditions for system (6.1) to define the endogenous variables $y_1, \ldots, y_m$ as differentiable functions of the exogenous variables $x_1, \ldots, x_n$. (For the case $n = m = 1$, see (4.17).)

6.4
$$f_1(x_1, x_2, \ldots, x_n) = 0$$
$$f_2(x_1, x_2, \ldots, x_n) = 0$$
$$\cdots\cdots\cdots\cdots\cdots\cdots\cdots$$
$$f_m(x_1, x_2, \ldots, x_n) = 0$$

A general system of m equations and n variables.

6.5 System (6.4) has k *degrees of freedom* if there is a set of k of the variables that can be freely chosen such that the remaining $n - k$ variables are uniquely determined when the k variables have been assigned specific values. If the variables are restricted to vary in a set S in $\mathbb{R}^n$, the system has k *degrees of freedom in S*.

Definition of degrees of freedom for a system of equations.

6.6 To find the number of degrees of freedom for a system of equations, count the number, n, of variables and the number, m, of equations. If $n > m$, there are $n - m$ degrees of freedom in the system. If $n < m$, there is, in general, no solution of the system.

The "counting rule". This is a rough rule which is *not* valid in general.

6.7 If the conditions in (6.3) are satisfied, then system (6.1) has n degrees of freedom.

A precise (local) counting rule.

6.8 $$\mathbf{f}'(\mathbf{x}) = \begin{pmatrix} \dfrac{\partial f_1(\mathbf{x})}{\partial x_1} & \cdots & \dfrac{\partial f_1(\mathbf{x})}{\partial x_n} \\ \vdots & & \vdots \\ \dfrac{\partial f_m(\mathbf{x})}{\partial x_1} & \cdots & \dfrac{\partial f_m(\mathbf{x})}{\partial x_n} \end{pmatrix}$$

The *Jacobian matrix* of $f_1, \ldots, f_m$ with respect to $x_1, \ldots, x_n$, also denoted by $\partial \mathbf{f}(\mathbf{x})/\partial \mathbf{x}$.

6.9 If $\mathbf{x}^0 = (x_1^0, \ldots, x_n^0)$ is a solution of (6.4), $m \leq n$, and the rank of the Jacobian matrix $\mathbf{f}'(\mathbf{x})$ is equal to m, then system (6.4) has $n - m$ degrees of freedom in some neighborhood of $\mathbf{x}^0$.

A precise (local) counting rule. (Valid if the functions $f_1, \ldots, f_m$ are C^1.)

6.10 The functions $f_1(\mathbf{x}), \ldots, f_m(\mathbf{x})$ are *functionally dependent* in an open set A in $\mathbb{R}^n$ if there exists a real-valued C^1 function F defined on an open set containing

$$S = \{(f_1(\mathbf{x}), \ldots, f_m(\mathbf{x})) : \mathbf{x} \in A\}$$

such that

$$F(f_1(\mathbf{x}), \ldots, f_m(\mathbf{x})) = 0 \quad \text{for all } \mathbf{x} \text{ in } A$$

and $\nabla F \neq \mathbf{0}$ in S.

Definition of functional dependence.

6.11 If $f_1(\mathbf{x}), \ldots, f_m(\mathbf{x})$ are functionally dependent in an open set $A \subset \mathbb{R}^n$, then the rank of the Jacobian matrix $\mathbf{f}'(\mathbf{x})$ is less than m for all $\mathbf{x}$ in A.

A necessary condition for functional dependence.

6.12 If the equation system (6.4) has solutions, and if $f_1(\mathbf{x}), \ldots, f_m(\mathbf{x})$ are functionally dependent, then (6.4) has at least one redundant equation.

A sufficient condition for the counting rule to fail.

6.13 $$\det(\mathbf{f}'(\mathbf{x})) = \begin{vmatrix} \dfrac{\partial f_1(\mathbf{x})}{\partial x_1} & \cdots & \dfrac{\partial f_1(\mathbf{x})}{\partial x_n} \\ \vdots & \ddots & \vdots \\ \dfrac{\partial f_n(\mathbf{x})}{\partial x_1} & \cdots & \dfrac{\partial f_n(\mathbf{x})}{\partial x_n} \end{vmatrix}$$

The *Jacobian determinant* of $f_1, \ldots, f_n$ with respect to $x_1, \ldots, x_n$. (See Chapter 20 for determinants.)

6.14 If $f_1(\mathbf{x}), \ldots, f_n(\mathbf{x})$ are functionally dependent, then the determinant $\det(\mathbf{f}'(\mathbf{x})) \equiv 0$.

A special case of (6.11). The converse is not generally true.

6.15 $$\begin{aligned} y_1 &= f_1(x_1, \ldots, x_n) \\ &\cdots\cdots\cdots\cdots\cdots \\ y_n &= f_n(x_1, \ldots, x_n) \end{aligned} \quad \Longleftrightarrow \quad \mathbf{y} = \mathbf{f}(\mathbf{x})$$

A *transformation* $\mathbf{f}$ from $\mathbb{R}^n$ to $\mathbb{R}^n$.

6.16 Suppose the transformation $\mathbf{f}$ in (6.15) is C^1 in a neighborhood of $\mathbf{x}^0$ and that the Jacobian determinant in (6.13) is not zero at $\mathbf{x} = \mathbf{x}^0$. Then there exists a C^1 transformation $\mathbf{g}$ that is locally an inverse of $\mathbf{f}$, i.e. $\mathbf{g}(\mathbf{f}(\mathbf{x})) = \mathbf{x}$ for all $\mathbf{x}$ in a neighborhood of $\mathbf{x}^0$ and $\mathbf{f}(\mathbf{g}(\mathbf{y})) = \mathbf{y}$ for all $\mathbf{y}$ in a neighborhood of $\mathbf{y}^0 = \mathbf{f}(\mathbf{x}^0)$.

The existence of a *local inverse*. (*Inverse function theorem*. Local version.)

6.17 Suppose $\mathbf{f} : \mathbb{R}^n \to \mathbb{R}^n$ is C^1 and that there exist positive numbers h and k such that
$$|\det(\mathbf{f}'(\mathbf{x}))| \geq h \text{ and } |\partial f_i(\mathbf{x})/\partial x_j| \leq k$$
for all $\mathbf{x}$ and all $i, j = 1, \ldots, n$. Then $\mathbf{f}$ has an inverse defined and C^1 on all of $\mathbb{R}^n$.

Existence of a *global inverse*. (Hadamard's theorem.)

6.18 Suppose $\mathbf{f} : \mathbb{R}^n \to \mathbb{R}^n$ is C^1 and that the determinant in (6.13) is $\neq 0$ for all $\mathbf{x}$. Then $\mathbf{f}(\mathbf{x})$ has an inverse that is C^1 and defined over all of $\mathbb{R}^n$, if and only if
$$\inf\{\|\mathbf{f}(\mathbf{x})\| : \|\mathbf{x}\| \geq n\} \to \infty \text{ as } n \to \infty.$$

A *global inverse function theorem*.

6.19 Suppose $\mathbf{f} : \mathbb{R}^n \to \mathbb{R}^n$ is C^1 and let Ω be the rectangle $\Omega = \{\mathbf{x} \in \mathbb{R}^n : \mathbf{a} \leq \mathbf{x} \leq \mathbf{b}\}$, where $\mathbf{a}$ and $\mathbf{b}$ are given vectors in $\mathbb{R}^n$. Then $\mathbf{f}$ is one-to-one in Ω if *one* of the following conditions is satisfied for all $\mathbf{x}$:

- The Jacobian matrix $\mathbf{f}'(\mathbf{x})$ has only strictly positive principal minors.
- The Jacobian matrix $\mathbf{f}'(\mathbf{x})$ has only strictly negative principal minors.

A *Gale–Nikaido theorem*. (For principal minors, see (20.15).)

6.20 An $n \times n$ matrix $\mathbf{A}$ (not necessarily symmetric) is called *positive quasidefinite* if $\mathbf{x}'\mathbf{A}\mathbf{x} > 0$ for every n-vector $\mathbf{x} \neq \mathbf{0}$.

Definition of a positive quasidefinite matrix.

6.21 Suppose $\mathbf{f} : \mathbb{R}^n \to \mathbb{R}^n$ is a C^1 function and assume that the Jacobian matrix $\mathbf{f}'(\mathbf{x})$ is positive quasidefinite everywhere in a convex set Ω. Then $\mathbf{f}$ is one-to-one in Ω.

A *Gale–Nikaido theorem*.

6.22 $\mathbf{f} : \mathbb{R}^n \to \mathbb{R}^n$ is called a *contraction mapping* if there exists a constant k in $[0, 1)$ such that
$$\|\mathbf{f}(\mathbf{x}) - \mathbf{f}(\mathbf{y})\| \leq k\|\mathbf{x} - \mathbf{y}\|$$
for all $\mathbf{x}$ and $\mathbf{y}$ in $\mathbb{R}^n$.

Definition of a contraction mapping.

6.23 If $\mathbf{f} : \mathbb{R}^n \to \mathbb{R}^n$ is a contraction mapping, then $\mathbf{f}$ has a unique *fixed point*, i.e. a point $\mathbf{x}^*$ in $\mathbb{R}^n$ such that $\mathbf{f}(\mathbf{x}^*) = \mathbf{x}^*$. For any $\mathbf{x}_0$ in $\mathbb{R}^n$ we have $\mathbf{x}^* = \lim_{n \to \infty} \mathbf{x}_n$, where $\mathbf{x}_n = \mathbf{f}(\mathbf{x}_{n-1})$ for $n \geq 1$.

The existence of a fixed point for a contraction mapping. (This result can be generalized to complete metric spaces. See (18.26).)

6.24 Let S be a nonempty subset of $\mathbb{R}^n$, and let $\mathcal{B}$ denote the set of all bounded functions from S into $\mathbb{R}^m$. The *supremum distance* between two functions φ and ψ in $\mathcal{B}$ is defined as
$$d(\varphi, \psi) = \sup_{\mathbf{x} \in S} \|\varphi(\mathbf{x}) - \psi(\mathbf{x})\|$$

A definition of distance between functions. ($F : S \to \mathbb{R}^m$ is called *bounded* on S if there exists a positive number M such that $\|F(\mathbf{x})\| \leq M$ for all $\mathbf{x}$ in S.)

6.25 Let S be a nonempty subset of $\mathbb{R}^n$ and let $\mathcal{B}$ be the set of all bounded functions from S into $\mathbb{R}^m$. Suppose that the function $T : \mathcal{B} \to \mathcal{B}$ is a contraction mapping in the sense that
$$d(T(\varphi), T(\psi)) \leq \beta d(\varphi, \psi) \quad \text{for all } \varphi, \psi \text{ in } \mathcal{B}$$
Then there exists a unique function φ^* in $\mathcal{B}$ such that $\varphi^* = T(\varphi^*)$.

A contraction mapping theorem for spaces of bounded functions.

6.26 Let K be a nonempty, compact and convex set in $\mathbb{R}^n$ and $\mathbf{f}$ a continuous function mapping K into K. Then $\mathbf{f}$ has a fixed point $\mathbf{x}^* \in K$, i.e. a point $\mathbf{x}^*$ such that $\mathbf{f}(\mathbf{x}^*) = \mathbf{x}^*$.

Brouwer's fixed point theorem.

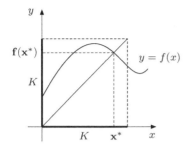

6.27

Illustration of Brouwer's fixed point theorem for $n = 1$.

6.28 Let K be a nonempty compact, convex set in $\mathbb{R}^n$ and $\mathbf{f}$ a correspondence that to each point $\mathbf{x}$ in K associates a nonempty, convex subset $\mathbf{f}(\mathbf{x})$ of K. Suppose that $\mathbf{f}$ has a closed graph, i.e. the set

$$\{(\mathbf{x}, \mathbf{y}) \in \mathbb{R}^{2n} : \mathbf{x} \in K \text{ and } \mathbf{y} \in \mathbf{f}(\mathbf{x})\}$$

is closed in $\mathbb{R}^{2n}$. Then $\mathbf{f}$ has a fixed point, i.e. a point $\mathbf{x}^*$ in K, such that $\mathbf{x}^* \in \mathbf{f}(\mathbf{x}^*)$.

Kakutani's fixed point theorem. (See (12.27) for the definition of correspondences.)

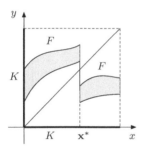

6.29

Illustration of Kakutani's fixed point theorem for $n = 1$.

6.30 If $\mathbf{x} = (x_1, \ldots, x_n)$ and $\mathbf{y} = (y_1, \ldots, y_n)$ are two points in $\mathbb{R}^n$, then the *meet* $\mathbf{x} \wedge \mathbf{y}$ and *join* $\mathbf{x} \vee \mathbf{y}$ of $\mathbf{x}$ and $\mathbf{y}$ are points in $\mathbb{R}^n$ defined as follows:

$$\mathbf{x} \wedge \mathbf{y} = (\min\{x_1, y_1\}, \ldots, \min\{x_n, y_n\})$$
$$\mathbf{x} \vee \mathbf{y} = (\max\{x_1, y_1\}, \ldots, \max\{x_n, y_n\})$$

Definition of the meet and the join of two vectors in $\mathbb{R}^n$.

6.31 A set S in $\mathbb{R}^n$ is called a *sublattice* of $\mathbb{R}^n$ if the meet and the join of any two points in S are also in S. If S is also a compact set, the S is called a *compact sublattice*.

Definition of a (compact) sublattice of $\mathbb{R}^n$.

6.32 Let S be a nonempty compact sublattice of $\mathbb{R}^n$. Let $\mathbf{f} : S \to S$ be an increasing function, i.e. if $\mathbf{x}, \mathbf{y} \in S$ and $\mathbf{x} \leq \mathbf{y}$, then $\mathbf{f}(\mathbf{x}) \leq \mathbf{f}(\mathbf{y})$. Then $\mathbf{f}$ has a fixed point in S, i.e. a point $\mathbf{x}^*$ in S such that $\mathbf{f}(\mathbf{x}^*) = \mathbf{x}^*$.

Tarski's fixed point theorem. (The theorem is not valid for decreasing functions. See (6.33).)

44

6.33

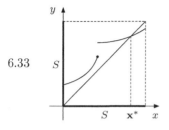 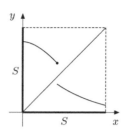

x^* is a fixed point for the increasing function in the figure to the left. The decreasing function in the other figure has no fixed point.

6.34

$$a_{11}x_1 + a_{12}x_2 + \cdots + a_{1n}x_n = b_1$$
$$a_{21}x_1 + a_{22}x_2 + \cdots + a_{2n}x_n = b_2$$
$$\cdots\cdots\cdots\cdots\cdots\cdots\cdots\cdots\cdots\cdots\cdots$$
$$a_{m1}x_1 + a_{m2}x_2 + \cdots + a_{mn}x_n = b_m$$

The *general linear system with m equations and n unknowns.*

6.35

$$\mathbf{A} = \begin{pmatrix} a_{11} & \cdots & a_{1n} \\ a_{21} & \cdots & a_{2n} \\ \vdots & & \vdots \\ a_{m1} & \cdots & a_{mn} \end{pmatrix}$$

$$\mathbf{A_b} = \begin{pmatrix} a_{11} & \cdots & a_{1n} & b_1 \\ a_{21} & \cdots & a_{2n} & b_2 \\ \vdots & & \vdots & \vdots \\ a_{m1} & \cdots & a_{mn} & b_m \end{pmatrix}$$

$\mathbf{A}$ is the coefficient matrix of (6.34), and $\mathbf{A_b}$ is the *augmented coefficient matrix.*

6.36

- System (6.34) has at least one solution if and only if $r(\mathbf{A}) = r(\mathbf{A_b})$.
- If $r(\mathbf{A}) = r(\mathbf{A_b}) = k < m$, then system (6.34) has $m - k$ superfluous equations.
- If $r(\mathbf{A}) = r(\mathbf{A_b}) = k < n$, then system (6.34) has $n - k$ degrees of freedom.

Main results about linear systems of equations. $r(\mathbf{B})$ denotes the rank of the matrix $\mathbf{B}$. (See (19.23).)

6.37

$$a_{11}x_1 + a_{12}x_2 + \cdots + a_{1n}x_n = 0$$
$$a_{21}x_1 + a_{22}x_2 + \cdots + a_{2n}x_n = 0$$
$$\cdots\cdots\cdots\cdots\cdots\cdots\cdots\cdots\cdots\cdots\cdots$$
$$a_{m1}x_1 + a_{m2}x_2 + \cdots + a_{mn}x_n = 0$$

The general *homogeneous* linear equation system with m equations and n unknowns.

6.38

- The homogeneous system (6.37) has a nontrivial solution if and only if $r(\mathbf{A}) < n$.
- If $n = m$, then the homogeneous system (6.37) has nontrivial solutions if and only if $|\mathbf{A}| = 0$.

Important results on homogeneous linear systems.

References

For (6.1)–(6.16) and (6.22)–(6.25), see e.g. Rudin (1982), Marsden and Hoffman (1993) or Sydsæter et al. (2005). For (6.17)–(6.21) see Parthasarathy (1983). For Brouwer's and Kakutani's fixed point theorems, see Nikaido (1970) or Scarf (1973). For Tarski's fixed point theorem and related material, see Sundaram (1996). (6.36)–(6.38) are standard results in linear algebra, see e.g. Fraleigh and Beauregard (1995), Lang (1987) or Sydsæter et al. (2005).

Chapter 7

Inequalities

7.1 $\left| |a| - |b| \right| \leq |a \pm b| \leq |a| + |b|$

Triangle inequalities.
$a, b \in \mathbb{R}$ (or $\mathbb{C}$).

7.2 $\dfrac{n}{\sum_{i=1}^{n} 1/a_i} \leq \left(\prod_{i=1}^{n} a_i \right)^{1/n} \leq \dfrac{\sum_{i=1}^{n} a_i}{n}, \quad a_i > 0$

Harmonic mean $\leq$
geometric mean $\leq$
arithmetic mean.
Equalities if and only if
$a_1 = \cdots = a_n$.

7.3 $\dfrac{2}{1/a_1 + 1/a_2} \leq \sqrt{a_1 a_2} \leq \dfrac{a_1 + a_2}{2}$

(7.2) for $n = 2$.

7.4 $(1 + x)^n \geq 1 + nx \quad (n \in \mathbb{N},\, x \geq -1)$

Bernoulli's inequality.

7.5 $a_1^{\lambda_1} \cdots a_n^{\lambda_n} \leq \lambda_1 a_1 + \cdots + \lambda_n a_n$

Inequality for *weighted*
means. $a_i \geq 0$,
$\sum_{i=1}^{n} \lambda_i = 1,\, \lambda_i \geq 0$.

7.6 $a_1^{\lambda} a_2^{1-\lambda} \leq \lambda a_1 + (1 - \lambda) a_2$

(7.5) for $n = 2$, $a_1 \geq 0$,
$a_2 \geq 0$, $\lambda \in [0, 1]$.

7.7 $\displaystyle\sum_{i=1}^{n} |a_i b_i| \leq \left[\sum_{i=1}^{n} |a_i|^p \right]^{1/p} \left[\sum_{i=1}^{n} |b_i|^q \right]^{1/q}$

Hölder's inequality.
$p, q > 1$, $1/p + 1/q = 1$.
Equality if $|b_i| = c|a_i|^{p-1}$
for a nonnegative con-
stant c.

7.8 $\left[\displaystyle\sum_{i=1}^{n} |a_i b_i| \right]^2 \leq \left[\sum_{i=1}^{n} a_i^2 \right] \left[\sum_{i=1}^{n} b_i^2 \right]$

Cauchy–Schwarz's in-
equality. (Put $p = q = 2$
in (7.7).)

7.9 $\left[\displaystyle\sum_{i=1}^{n} a_i \right] \left[\sum_{i=1}^{n} b_i \right] \leq n \sum_{i=1}^{n} a_i b_i$

Chebyshev's inequality.
$a_1 \geq \cdots \geq a_n$,
$b_1 \geq \cdots \geq b_n$.

7.10	$$\left[\sum_{i=1}^{n}|a_i+b_i|^p\right]^{1/p} \leq \left[\sum_{i=1}^{n}|a_i|^p\right]^{1/p} + \left[\sum_{i=1}^{n}|b_i|^p\right]^{1/p}$$	*Minkowski's inequality.* $p \geq 1$. Equality if $b_i = ca_i$ for a nonnegative constant c.
7.11	If f is convex, then $f\left[\sum_{i=1}^{n}a_i x_i\right] \leq \sum_{i=1}^{n}a_i f(x_i)$	*Jensen's inequality.* $\sum_{i=1}^{n}a_i = 1$, $a_i \geq 0$, $i = 1, \ldots, n$.
7.12	$$\left[\sum_{i=1}^{n}|a_i|^q\right]^{1/q} \leq \left[\sum_{i=1}^{n}|a_i|^p\right]^{1/p}$$	Another *Jensen's inequality*; $0 < p < q$.
7.13	$$\int_a^b |f(x)g(x)|\,dx \leq$$ $$\left[\int_a^b |f(x)|^p\,dx\right]^{1/p}\left[\int_a^b |g(x)|^q\,dx\right]^{1/q}$$	*Hölder's inequality.* $p > 1$, $q > 1$, $1/p + 1/q = 1$. Equality if $|g(x)| = c|f(x)|^{p-1}$ for a nonnegative constant c.
7.14	$$\left[\int_a^b f(x)g(x)\,dx\right]^2 \leq \int_a^b (f(x))^2\,dx \int_a^b (g(x))^2\,dx$$	*Cauchy–Schwarz's* inequality.
7.15	$$\left[\int_a^b |f(x)+g(x)|^p\,dx\right]^{1/p} \leq$$ $$\left[\int_a^b |f(x)|^p\,dx\right]^{1/p} + \left[\int_a^b |g(x)|^p\,dx\right]^{1/p}$$	*Minkowski's inequality.* $p \geq 1$. Equality if $g(x) = cf(x)$ for a nonnegative constant c.
7.16	If f is convex, then $$f\left(\int a(x)g(x)\,dx\right) \leq \int a(x)f(g(x))\,dx$$	*Jensen's inequality.* $a(x) \geq 0$, $f(u) \geq 0$, $\int a(x)\,dx = 1$. f is defined on the range of g.
7.17	If f is convex on the interval I and X is a random variable with finite expectation, then $$f(E[X]) \leq E[f(X)]$$ If f is strictly convex, the inequality is strict unless X is a constant with probability 1.	Special case of Jensen's inequality. E is the expectation operator.
7.18	If U is concave on the interval I and X is a random variable with finite expectation, then $$E[U(X)] \leq U(E[X])$$	An important fact in utility theory. (It follows from (7.17) by putting $f = -U$.)

References

Hardy, Littlewood, and Pólya (1952) is still a good reference for inequalities.

Chapter 8

Series. Taylor's formula

<table>
<tr><td>8.1</td><td>$\displaystyle\sum_{i=0}^{n-1}(a + id) = na + \frac{n(n-1)d}{2}$</td><td>Sum of the first n terms of an *arithmetic series*.</td></tr>
<tr><td>8.2</td><td>$a + ak + ak^2 + \cdots + ak^{n-1} = a\,\dfrac{1 - k^n}{1 - k},\ \ k \neq 1$</td><td>Sum of the first n terms of a *geometric series*.</td></tr>
<tr><td>8.3</td><td>$a + ak + \cdots + ak^{n-1} + \cdots = \dfrac{a}{1 - k}\ \ \text{if } |k| < 1$</td><td>Sum of an infinite geometric series.</td></tr>
<tr><td>8.4</td><td>$\displaystyle\sum_{n=1}^{\infty} a_n = s\quad\text{means that}\quad \lim_{n\to\infty}\sum_{k=1}^{n} a_k = s$</td><td>Definition of the *convergence* of an infinite series. If the series does not converge, it *diverges*.</td></tr>
<tr><td>8.5</td><td>$\displaystyle\sum_{n=1}^{\infty} a_n \text{ converges}\ \Rightarrow\ \lim_{n\to\infty} a_n = 0$</td><td>A necessary (but NOT sufficient) condition for the convergence of an infinite series.</td></tr>
<tr><td>8.6</td><td>$\displaystyle\lim_{n\to\infty}\left|\frac{a_{n+1}}{a_n}\right| < 1\ \Rightarrow\ \sum_{n=1}^{\infty} a_n \text{ converges}$</td><td>The *ratio test*.</td></tr>
<tr><td>8.7</td><td>$\displaystyle\lim_{n\to\infty}\left|\frac{a_{n+1}}{a_n}\right| > 1\ \Rightarrow\ \sum_{n=1}^{\infty} a_n \text{ diverges}$</td><td>The *ratio test*.</td></tr>
<tr><td>8.8</td><td>If $f(x)$ is a positive-valued, decreasing, and continuous function for $x \geq 1$, and if $a_n = f(n)$ for all integers $n \geq 1$, then the infinite series and the improper integral

$\displaystyle\sum_{n=1}^{\infty} a_n\quad\text{and}\quad \int_1^{\infty} f(x)\,dx$

either both converge or both diverge.</td><td>The *integral test*.</td></tr>
</table>

If $0 \leq a_n \leq b_n$ for all n, then

8.9
- $\sum a_n$ converges if $\sum b_n$ converges.
- $\sum b_n$ diverges if $\sum a_n$ diverges.

| The *comparison test.*

8.10 $\displaystyle\sum_{n=1}^{\infty} \frac{1}{n^p}$ is convergent $\iff$ $p > 1$

| An important result.

8.11 A series $\displaystyle\sum_{n=1}^{\infty} a_n$ is said to *converge absolutely* if the series $\displaystyle\sum_{n=1}^{\infty} |a_n|$ converges.

| Definition of absolute convergence. $|a_n|$ denotes the absolute value of a_n.

8.12 Every absolutely convergent series is convergent, but not all convergent series are absolutely convergent.

| A convergent series that is not absolutely convergent, is called *conditionally convergent.*

8.13 If a series is absolutely convergent, then the sum is independent of the order in which terms are summed. A conditionally convergent series can be made to converge to any number (or even diverge) by suitable rearranging the order of the terms.

| Important results on the convergence of series.

8.14 $f(x) \approx f(a) + f'(a)(x-a)$ (x close to a)

| *First-order (linear)* approximation about $x = a$.

8.15 $f(x) \approx f(a) + f'(a)(x-a) + \frac{1}{2}f''(a)(x-a)^2$
(x close to a)

| *Second-order (quadratic)* approximation about $x = a$.

8.16 $f(x) = f(0) + \dfrac{f'(0)}{1!}x + \cdots + \dfrac{f^{(n)}(0)}{n!}x^n$
$\quad + \dfrac{f^{(n+1)}(\theta x)}{(n+1)!}x^{n+1}, \quad 0 < \theta < 1$

| *Maclaurin's formula.* The last term is Lagrange's error term.

8.17 $f(x) = f(0) + \dfrac{f'(0)}{1!}x + \dfrac{f''(0)}{2!}x^2 + \cdots$

| The *Maclaurin series* for $f(x)$, valid for those x for which the error term in (8.16) tends to 0 as n tends to ∞.

8.18 $f(x) = f(a) + \dfrac{f'(a)}{1!}(x-a) + \cdots + \dfrac{f^{(n)}(a)}{n!}(x-a)^n$
$\quad + \dfrac{f^{(n+1)}(a + \theta(x-a))}{(n+1)!}(x-a)^{n+1}, \quad 0 < \theta < 1$

| *Taylor's formula.* The last term is Lagrange's error term.

8.19 $\quad f(x) = f(a) + \dfrac{f'(a)}{1!}(x-a) + \dfrac{f''(a)}{2!}(x-a)^2 + \cdots$

The *Taylor series* for $f(x)$, valid for those x where the error term in (8.18) tends to 0 as n tends to ∞.

8.20 $\quad f(x,y) \approx f(a,b) + f_1'(a,b)(x-a) + f_2'(a,b)(y-b)$
$$((x,y) \text{ close to } (a,b))$$

First-order (linear) approximation to $f(x,y)$ about (a,b).

8.21
$f(x,y) \approx$
$f(a,b) + f_1'(a,b)(x-a) + f_2'(a,b)(y-b)$
$+\frac{1}{2}[f_{11}''(a,b)(x-a)^2 + 2f_{12}''(a,b)(x-a)(y-b) + f_{22}''(a,b)(y-b)^2]$

Second-order (quadratic) approximation to $f(x,y)$ about (a,b).

8.22
$$f(\mathbf{x}) = f(\mathbf{a}) + \sum_{i=1}^{n} f_i'(\mathbf{a})(x_i - a_i)$$
$$+\frac{1}{2}\sum_{i=1}^{n}\sum_{j=1}^{n} f_{ij}''(\mathbf{a} + \theta(\mathbf{x} - \mathbf{a}))(x_i - a_i)(x_j - a_j)$$

Taylor's formula of order 2 for functions of n variables, $\theta \in (0,1)$.

8.23 $\quad e^x = 1 + \dfrac{x}{1!} + \dfrac{x^2}{2!} + \dfrac{x^3}{3!} + \cdots$

Valid for all x.

8.24 $\quad \ln(1+x) = x - \dfrac{x^2}{2} + \dfrac{x^3}{3} - \dfrac{x^4}{4} + \cdots$

Valid if $-1 < x \leq 1$.

8.25 $\quad (1+x)^m = \dbinom{m}{0} + \dbinom{m}{1}x + \dbinom{m}{2}x^2 + \cdots$

Valid if $-1 < x < 1$. For the definition of $\binom{m}{k}$, see (8.30).

8.26 $\quad \sin x = x - \dfrac{x^3}{3!} + \dfrac{x^5}{5!} - \dfrac{x^7}{7!} + \dfrac{x^9}{9!} - \cdots$

Valid for all x.

8.27 $\quad \cos x = 1 - \dfrac{x^2}{2!} + \dfrac{x^4}{4!} - \dfrac{x^6}{6!} + \dfrac{x^8}{8!} - \cdots$

Valid for all x.

8.28 $\quad \arcsin x = x + \dfrac{1}{2}\dfrac{x^3}{3} + \dfrac{1 \cdot 3}{2 \cdot 4}\dfrac{x^5}{5} + \dfrac{1 \cdot 3 \cdot 5}{2 \cdot 4 \cdot 6}\dfrac{x^7}{7} + \cdots$

Valid if $|x| \leq 1$.

8.29 $\quad \arctan x = x - \dfrac{x^3}{3} + \dfrac{x^5}{5} - \dfrac{x^7}{7} + \cdots$

Valid if $|x| \leq 1$.

8.30
- $\dbinom{r}{k} = \dfrac{r(r-1)\cdots(r-k+1)}{k!}$
- $\dbinom{r}{0} = 1, \qquad \dbinom{r}{-k} = 0$

Binomial coefficients. (r is an arbitrary real number, k is a natural number.)

8.31
- $\dbinom{n}{k} = \dfrac{n!}{k!(n-k)!}$ $(0 \le k \le n)$
- $\dbinom{n}{k} = \dbinom{n}{n-k}$ $(n \ge 0)$
- $\dbinom{r}{k} = \dbinom{r-1}{k-1} + \dbinom{r-1}{k}$
- $\dbinom{-r}{k} = (-1)^k \dbinom{r+k-1}{k}$
- $\dbinom{r}{k} = \dfrac{r}{k}\dbinom{r-1}{k-1}$ $(k \ne 0)$

Important properties of the binomial coefficients. n and k are integers, and r is a real number.

8.32 $\dbinom{r}{m}\dbinom{m}{k} = \dbinom{r}{k}\dbinom{r-k}{m-k}$

m and k are integers.

8.33 $\displaystyle\sum_{k=0}^{n} \dbinom{r}{k}\dbinom{s}{n-k} = \dbinom{r+s}{n}$

n is a nonnegative integer.

8.34 $\displaystyle\sum_{k=0}^{n} \dbinom{r+k}{k} = \dbinom{r+n+1}{n}$

n is a nonnegative integer.

8.35 $\displaystyle\sum_{k=0}^{n} \dbinom{k}{m} = \dbinom{n+1}{m+1}$

m and n are nonnegative integers.

8.36 $\displaystyle\sum_{k=0}^{n} \dbinom{n}{k} a^{n-k}b^k = (a+b)^n$

Newton's binomial formula.

8.37 $\displaystyle\sum_{k=0}^{n} \dbinom{n}{k} = (1+1)^n = 2^n$

A special case of (8.36).

8.38 $\displaystyle\sum_{k=0}^{n} \dbinom{n}{k} k = n2^{n-1}$ $(n \ge 0)$

8.39 $\displaystyle\sum_{k=0}^{n} \dbinom{n}{k} k^2 = (n^2+n)2^{n-2}$ $(n \ge 0)$

8.40 $\displaystyle\sum_{k=0}^{n} \dbinom{n}{k}^2 = \dbinom{2n}{n}$

8.41 $(a_1 + a_2 + \cdots + a_m)^n =$
$$\sum_{k_1+\cdots+k_m=n} \frac{n!}{k_1! \cdots k_m!} a_1^{k_1} \cdots a_m^{k_m}$$

The *multinomial formula*.

8.42 $1 + 2 + 3 + \cdots + n = \dfrac{n(n+1)}{2}$ | Summation formulas.

8.43 $1 + 3 + 5 + \cdots + (2n - 1) = n^2$ |

8.44 $1^2 + 2^2 + 3^2 + \cdots + n^2 = \dfrac{n(n+1)(2n+1)}{6}$ |

8.45 $1^3 + 2^3 + 3^3 + \cdots + n^3 = \left(\dfrac{n(n+1)}{2}\right)^2$ |

8.46 $$1^4 + 2^4 + 3^4 + \cdots + n^4 = \dfrac{n(n+1)(2n+1)(3n^2 + 3n - 1)}{30}$$ |

8.47 $\dfrac{1}{1^2} + \dfrac{1}{2^2} + \dfrac{1}{3^2} + \cdots + \dfrac{1}{n^2} + \cdots = \dfrac{\pi^2}{6}$ | A famous result.

8.48 $\displaystyle\lim_{n \to \infty} \left[\left(\dfrac{1}{1} + \dfrac{1}{2} + \cdots + \dfrac{1}{n}\right) - \ln n\right] = \gamma \approx 0.5772\ldots$ | The constant γ is called *Euler's constant*.

References

All formulas are standard and are usually found in calculus texts, e.g. Edwards and Penney (1998). For results about binomial coefficients, see a book on probability theory, or e.g. Graham, Knuth, and Patashnik (1989).

Chapter 9

Integration

Indefinite integrals

9.1 $\displaystyle\int f(x)\,dx = F(x) + C \iff F'(x) = f(x)$

Definition of the *indefinite integral*.

9.2 $\displaystyle\int (af(x) + bg(x))\,dx = a\int f(x)\,dx + b\int g(x)\,dx$

Linearity of the integral. a and b are constants.

9.3 $\displaystyle\int f(x)g'(x)\,dx = f(x)g(x) - \int f'(x)g(x)\,dx$

Integration by parts.

9.4 $\displaystyle\int f(x)\,dx = \int f(g(t))g'(t)\,dt, \quad x = g(t)$

Change of variable. (Integration by substitution.)

9.5 $\displaystyle\int x^n\,dx = \begin{cases} \dfrac{x^{n+1}}{n+1} + C, & n \neq -1 \\[2mm] \ln|x| + C, & n = -1 \end{cases}$

Special integration results.

9.6 $\displaystyle\int a^x\,dx = \frac{1}{\ln a}a^x + C, \quad a > 0, \; a \neq 1$

9.7 $\displaystyle\int e^x\,dx = e^x + C$

9.8 $\displaystyle\int xe^x\,dx = xe^x - e^x + C$

9.9 $\displaystyle\int x^n e^{ax}\,dx = \frac{x^n}{a}e^{ax} - \frac{n}{a}\int x^{n-1}e^{ax}\,dx, \; a \neq 0$

9.10 $\displaystyle\int \log_a x\,dx = x\log_a x - x\log_a e + C, \quad a > 0, \; a \neq 1$

9.11 $\displaystyle\int \ln x \, dx = x \ln x - x + C$

Special integration results.

9.12 $\displaystyle\int x^n \ln x \, dx = \dfrac{x^{n+1}\big((n+1)\ln x - 1\big)}{(n+1)^2} + C$ $\qquad (n \neq -1)$

9.13 $\displaystyle\int \sin x \, dx = -\cos x + C$

9.14 $\displaystyle\int \cos x \, dx = \sin x + C$

9.15 $\displaystyle\int \tan x \, dx = -\ln|\cos x| + C$

9.16 $\displaystyle\int \cot x \, dx = \ln|\sin x| + C$

9.17 $\displaystyle\int \dfrac{1}{\sin x} \, dx = \ln\left|\dfrac{1 - \cos x}{\sin x}\right| + C$

9.18 $\displaystyle\int \dfrac{1}{\cos x} \, dx = \ln\left|\dfrac{1 + \sin x}{\cos x}\right| + C$

9.19 $\displaystyle\int \dfrac{1}{\sin^2 x} \, dx = -\cot x + C$

9.20 $\displaystyle\int \dfrac{1}{\cos^2 x} \, dx = \tan x + C$

9.21 $\displaystyle\int \sin^2 x \, dx = \dfrac{1}{2}x - \dfrac{1}{2}\sin x \cos x + C$

9.22 $\displaystyle\int \cos^2 x \, dx = \dfrac{1}{2}x + \dfrac{1}{2}\sin x \cos x + C$

9.23 $\displaystyle\int \sin^n x \, dx =$
$$-\dfrac{\sin^{n-1} x \cos x}{n} + \dfrac{n-1}{n}\int \sin^{n-2} x \, dx$$
$\qquad (n \neq 0)$

9.24 $\displaystyle\int \cos^n x \, dx =$
$$\dfrac{\cos^{n-1} x \sin x}{n} + \dfrac{n-1}{n}\int \cos^{n-2} x \, dx$$
$\qquad (n \neq 0)$

9.25
$$\int e^{\alpha x} \sin \beta x \, dx =$$
$$\frac{e^{\alpha x}}{\alpha^2 + \beta^2} (\alpha \sin \beta x - \beta \cos \beta x) + C$$
$\quad (\alpha^2 + \beta^2 \neq 0)$

9.26
$$\int e^{\alpha x} \cos \beta x \, dx =$$
$$\frac{e^{\alpha x}}{\alpha^2 + \beta^2} (\beta \sin \beta x + \alpha \cos \beta x) + C$$
$\quad (\alpha^2 + \beta^2 \neq 0)$

9.27
$$\int \frac{1}{x^2 - a^2} \, dx = \frac{1}{2a} \ln \left| \frac{x-a}{x+a} \right| + C$$
$\quad (a \neq 0)$

9.28
$$\int \frac{1}{x^2 + a^2} \, dx = \frac{1}{a} \arctan \frac{x}{a} + C$$
$\quad (a \neq 0)$

9.29
$$\int \frac{1}{\sqrt{a^2 - x^2}} \, dx = \arcsin \frac{x}{a} + C$$
$\quad (a > 0)$

9.30
$$\int \frac{1}{\sqrt{x^2 \pm a^2}} \, dx = \ln \left| x + \sqrt{x^2 \pm a^2} \right| + C$$

9.31
$$\int \sqrt{a^2 - x^2} \, dx = \frac{x}{2} \sqrt{a^2 - x^2} + \frac{a^2}{2} \arcsin \frac{x}{a} + C$$
$\quad (a > 0)$

9.32
$$\int \sqrt{x^2 \pm a^2} \, dx =$$
$$\frac{x}{2} \sqrt{x^2 \pm a^2} \pm \frac{a^2}{2} \ln \left| x + \sqrt{x^2 \pm a^2} \right| + C$$

9.33
$$\int \frac{dx}{ax^2 + 2bx + c} =$$
$$\frac{1}{2\sqrt{b^2 - ac}} \ln \left| \frac{ax + b - \sqrt{b^2 - ac}}{ax + b + \sqrt{b^2 - ac}} \right| + C$$
$\quad (b^2 > ac, \ a \neq 0)$

9.34
$$\int \frac{dx}{ax^2 + 2bx + c} =$$
$$\frac{1}{\sqrt{ac - b^2}} \arctan \frac{ax + b}{\sqrt{ac - b^2}} + C$$
$\quad (b^2 < ac)$

9.35
$$\int \frac{dx}{ax^2 + 2bx + c} = \frac{-1}{ax + b} + C$$
$\quad (b^2 = ac, \ a \neq 0)$

$$\int \frac{x}{1+x^2}\,dx = \frac{1}{2}\int \frac{du}{u} = \frac{1}{2}\ln|u| + c = \frac{1}{2}\ln(1+x^2) + c$$

58

Let $u = 1+x^2$
then $du = 2x\,dx$

EX $\int \frac{x}{\sqrt{1+x^2}}\,dx = \frac{1}{2}\int \frac{du}{\sqrt{u}} = \frac{1}{2}\int u^{-1/2}\,du$

Definite integrals

$$= \frac{1}{2}\frac{u^{1/2}}{1/2} + c = (1+x^2)^{1/2} + c$$

9.36	$\displaystyle\int_a^b f(x)\,dx = \Big|_a^b F(x) = F(b) - F(a)$ if $F'(x) = f(x)$ for all x in $[a,b]$.	Definition of the *definite integral* of a function f.
9.37	• $A(x) = \displaystyle\int_a^x f(t)\,dt \;\Rightarrow\; A'(x) = f(x)$ • $A(x) = \displaystyle\int_x^b f(t)\,dt \;\Rightarrow\; A'(x) = -f(x)$	Important facts.
9.38		The shaded area is $A(x) = \int_a^x f(t)\,dt$, and the derivative of the area function $A(x)$ is $A'(x) = f(x)$.
9.39	$\displaystyle\int_a^b f(x)\,dx = -\int_b^a f(x)\,dx$ $\displaystyle\int_a^a f(x)\,dx = 0$ $\displaystyle\int_a^b \alpha f(x)\,dx = \alpha\int_a^b f(x)\,dx$ $\displaystyle\int_a^b f(x)\,dx = \int_a^c f(x)\,dx + \int_c^b f(x)\,dx$	a, b, c, and α are arbitrary real numbers.
9.40	$\displaystyle\int_a^b f(g(x))g'(x)\,dx = \int_{g(a)}^{g(b)} f(u)\,du, \quad u = g(x)$	*Change of variable. (Integration by substitution.)*
9.41	$\displaystyle\int_a^b f(x)g'(x)\,dx = \Big|_a^b f(x)g(x) - \int_a^b f'(x)g(x)\,dx$	*Integration by parts.*
9.42	$\displaystyle\int_a^\infty f(x)\,dx = \lim_{M\to\infty}\int_a^M f(x)\,dx$	If the limit exists, the integral is *convergent*. (In the opposite case, the integral *diverges*.)
9.43	$\displaystyle\int_{-\infty}^b f(x)\,dx = \lim_{N\to\infty}\int_{-N}^b f(x)\,dx$	If the limit exists, the integral is *convergent*. (In the opposite case, the integral *diverges*.)

9.44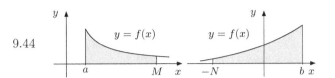

The figures illustrate (9.42) and (9.43). The shaded areas are $\int_a^M f(x)\,dx$ in the first figure, and $\int_{-N}^b f(x)\,dx$ in the second.

9.45
$$\int_{-\infty}^{\infty} f(x)\,dx = \int_{-\infty}^{a} f(x)\,dx + \int_{a}^{\infty} f(x)\,dx$$
$$= \lim_{N\to\infty} \int_{-N}^{a} f(x)\,dx + \lim_{M\to\infty} \int_{a}^{M} f(x)\,dx$$

Both limits on the right-hand side must exist. a is an arbitrary number. The integral is then said to *converge*. (If either of the limits does not exist, the integral *diverges*.)

9.46
$$\int_{a}^{b} f(x)\,dx = \lim_{h\to 0^+} \int_{a+h}^{b} f(x)\,dx$$

The definition of the integral if f is continuous in $(a, b]$.

9.47
$$\int_{a}^{b} f(x)\,dx = \lim_{h\to 0^+} \int_{a}^{b-h} f(x)\,dx$$

The definition of the integral if f is continuous in $[a, b)$.

9.48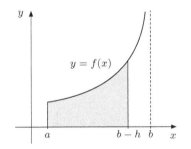

Illustrating definition (9.47). The shaded area is $\int_a^{b-h} f(x)\,dx$.

9.49
$$|f(x)| \le g(x) \text{ for all } x \ge a \implies$$
$$\left| \int_{a}^{\infty} f(x)\,dx \right| \le \int_{a}^{\infty} g(x)\,dx$$

Comparison test for integrals. f and g are continuous for $x \ge a$.

9.50
$$\frac{d}{dx} \int_{a}^{b} f(x,t)\,dt = \int_{a}^{b} f_x'(x,t)\,dt$$

"Differentiation under the integral sign". a and b are independent of x.

9.51
$$\frac{d}{dx} \int_{c}^{\infty} f(x,t)\,dt = \int_{c}^{\infty} f_x'(x,t)\,dt$$

Valid for x in (a, b) if $f(x,t)$ and $f_x'(x,t)$ are continuous for all $t \ge c$ and all x in (a, b), and $\int_c^{\infty} f(x,t)\,dt$ and $\int_c^{\infty} f_x'(x,t)\,dt$ converge uniformly on (a, b).

9.52
$$\frac{d}{dx}\int_{u(x)}^{v(x)} f(x,t)\,dt =$$

$$f\big(x,v(x)\big)v'(x) - f\big(x,u(x)\big)u'(x) + \int_{u(x)}^{v(x)} f'_x(x,t)\,dt$$

Leibniz's formula.

9.53 $\Gamma(x) = \displaystyle\int_0^\infty e^{-t}t^{x-1}\,dt, \quad x > 0$

The *gamma function.*

9.54
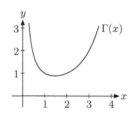

The graph of the gamma function. The minimum value is ≈ 0.8856 at $x \approx 1.4616$.

9.55 $\Gamma(x+1) = x\,\Gamma(x) \quad$ for all $x > 0$

The *functional equation* for the gamma function.

9.56 $\Gamma(n) = (n-1)! \quad$ when n is a positive integer.

Follows immediately from the functional equation.

9.57 $\displaystyle\int_{-\infty}^{+\infty} e^{-at^2}\,dt = \sqrt{\pi/a} \qquad (a > 0)$

An important formula.

9.58
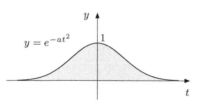

According to (9.57) the shaded area is $\sqrt{\pi/a}$.

9.59 $\displaystyle\int_0^\infty t^k e^{-at^2}\,dt = \frac{1}{2}a^{-(k+1)/2}\Gamma((k+1)/2)$

Valid for $a > 0$, $k > -1$.

9.60 $\Gamma(x) = \sqrt{2\pi}\,x^{x-\frac{1}{2}}e^{-x}e^{\theta/12x}, \quad x > 0,\ \theta \in (0,1)$

Stirling's formula.

9.61 $B(p,q) = \displaystyle\int_0^1 u^{p-1}(1-u)^{q-1}\,du, \quad p,\,q > 0$

The *beta function.*

9.62 $B(p,q) = \dfrac{\Gamma(p)\Gamma(q)}{\Gamma(p+q)}$

The relationship between the beta function and the gamma function.

9.63 $\displaystyle\int_a^b f(x)\,dx \approx \frac{b-a}{2n}[f(x_0) + 2\sum_{i=1}^{n-1} f(x_i) + f(x_n)]$

The *trapezoid formula*. $x_i = a + i\dfrac{b-a}{n}$, $i = 0,\dots,n$.

9.64 If f is C^2 on $[a,b]$ and $|f''(x)| \le M$ for all x in $[a,b]$, then $M(b-a)^3/12n^2$ is an upper bound on the error of approximation in (9.63).

Trapezoidal error estimate.

9.65 $\displaystyle\int_a^b f(x)\,dx \approx \frac{b-a}{6n}D,$ where $D =$

$$f(x_0) + 4\sum_{i=1}^{n} f(x_{2i-1}) + 2\sum_{i=1}^{n-1} f(x_{2i}) + f(x_{2n})$$

Simpson's formula. The points $x_j = a + j\dfrac{b-a}{2n}$, $j = 0,\dots,2n$, partition $[a,b]$ into $2n$ equal subintervals.

9.66 If f is C^4 on $[a,b]$ and $|f^{(4)}(x)| \le M$ for all x in $[a,b]$, then $M(b-a)^5/180n^4$ is an upper bound on the error of approximation in (9.65).

Simpson's error estimate.

Multiple integrals

9.67 $\displaystyle\iint_R f(x,y)\,dx\,dy = \int_a^b \left(\int_c^d f(x,y)\,dy\right) dx$

$$= \int_c^d \left(\int_a^b f(x,y)\,dx\right) dy$$

Definition of the *double integral* of $f(x,y)$ over a rectangle $R = [a,b] \times [c,d]$. (The fact that the two iterated integrals are equal for continuous functions, is *Fubini's theorem*.)

9.68 $\displaystyle\iint_{\Omega_A} f(x,y)\,dx\,dy = \int_a^b \left(\int_{u(x)}^{v(x)} f(x,y)\,dy\right) dx$

The double integral of a function $f(x,y)$ over the region Ω_A in figure A.

9.69 $\displaystyle\iint_{\Omega_B} f(x,y)\,dx\,dy = \int_c^d \left(\int_{p(y)}^{q(y)} f(x,y)\,dx\right) dy$

The double integral of a function $f(x,y)$ over the region Ω_B in figure B.

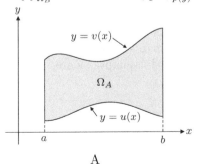

A

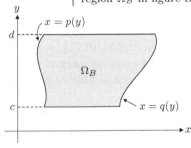

B

9.70
$$F''_{xy}(x,y) = f(x,y), \quad (x,y) \in [a,b] \times [c,d] \Rightarrow$$
$$\int_c^d \left(\int_a^b f(x,y)\, dx \right) dy =$$
$$F(b,d) - F(a,d) - F(b,c) + F(a,c)$$

An interesting result. $f(x,y)$ is a continuous function.

9.71
$$\iint_A f(x,y)\, dx\, dy =$$
$$\iint_{A'} f(g(u,v), h(u,v))|J|\, du\, dv$$

Change of variables in a double integral. $x = g(u,v)$, $y = h(u,v)$ is a one-to-one C^1 transformation of A' onto A, and the Jacobian determinant $J = \partial(g,h)/\partial(u,v)$ does not vanish in A'. f is continuous.

9.72
$$\iint \cdots \int_\Omega f(\mathbf{x})\, dx_1 \ldots dx_{n-1}\, dx_n =$$
$$\int_{a_n}^{b_n} \left(\int_{a_{n-1}}^{b_{n-1}} \cdots \left(\int_{a_1}^{b_1} f(\mathbf{x})\, dx_1 \right) \cdots dx_{n-1} \right) dx_n$$

The n-integral of f over an n-dimensional rectangle Ω. $\mathbf{x} = (x_1, \ldots, x_n)$.

9.73
$$\int \cdots \int_A f(\mathbf{x})\, dx_1 \ldots dx_n =$$
$$\int \cdots \int_{A'} f(g_1(\mathbf{u}), \ldots, g_n(\mathbf{u})) |J|\, du_1 \ldots du_n$$

Change of variables in the n-integral. $x_i = g_i(\mathbf{u})$, $i = 1, \ldots, n$, is a one-to-one C^1 transformation of A' onto A, and the Jacobian determinant
$$J = \frac{\partial(g_1, \ldots, g_n)}{\partial(u_1, \ldots, u_n)}$$
does not vanish in A'. f is continuous.

References

Most of these formulas can be found in any calculus text, e.g. Edwards and Penney (1998). For (9.67)–(9.73), see Marsden and Hoffman (1993), who have a precise treatment of multiple integrals. (Not all the required assumptions are spelled out in the subsection on multiple integrals.)

Chapter 10

Difference equations

10.1 $\quad x_t = a_t x_{t-1} + b_t, \qquad t = 1, 2, \ldots$

$\quad$ A *first-order linear difference equation.*

10.2 $\quad x_t = \left(\prod_{s=1}^{t} a_s \right) x_0 + \sum_{k=1}^{t} \left(\prod_{s=k+1}^{t} a_s \right) b_k$

$\quad$ The solution of (10.1) if we define the "empty" product $\prod_{s=t+1}^{t} a_s$ as 1.

10.3 $\quad x_t = a^t x_0 + \sum_{k=1}^{t} a^{t-k} b_k, \qquad t = 1, 2, \ldots$

$\quad$ The solution of (10.1) when $a_t = a$, a constant.

10.4
- $x_t = A a^t + \sum_{s=0}^{\infty} a^s b_{t-s}, \qquad |a| < 1$
- $x_t = A a^t - \sum_{s=1}^{\infty} \left(\frac{1}{a} \right)^s b_{t+s}, \qquad |a| > 1$

$\quad$ The *backward* and *forward solutions* of (10.1), respectively, with $a_t = a$, and with A as an arbitrary constant.

10.5 $\quad x_t = a x_{t-1} + b \iff x_t = a^t \left(x_0 - \frac{b}{1-a} \right) + \frac{b}{1-a}$

$\quad$ Equation (10.1) and its solution when $a_t = a \neq 1,\ b_t = b$.

10.6
$\quad (*) \quad x_t + a_1(t) x_{t-1} + \cdots + a_n(t) x_{t-n} = b_t$
$\quad (**) \quad x_t + a_1(t) x_{t-1} + \cdots + a_n(t) x_{t-n} = 0$

$\quad$ $(*)$ is the *general linear inhomogeneous difference equation of order n*, and $(**)$ is the associated *homogeneous* equation.

10.7
If $u_1(t), \ldots, u_n(t)$ are linearly independent solutions of (10.6) $(**)$, u_t^* is some particular solution of (10.6) $(*)$, and $C_1, \ldots, C_n$ are arbitrary constants, then the general solution of $(**)$ is

$$x_t = C_1 u_1(t) + \cdots + C_n u_n(t)$$

and the general solution of $(*)$ is

$$x_t = C_1 u_1(t) + \cdots + C_n u_n(t) + u_t^*$$

$\quad$ The structure of the solutions of (10.6). (For linear independence, see (11.21).)

For $b \neq 0$, $x_t + ax_{t-1} + bx_{t-2} = 0$ has the solution:

- For $\frac{1}{4}a^2 - b > 0$: $x_t = C_1 m_1^t + C_2 m_2^t$,

  where $m_{1,2} = -\frac{1}{2}a \pm \sqrt{\frac{1}{4}a^2 - b}$.

10.8

- For $\frac{1}{4}a^2 - b = 0$: $x_t = (C_1 + C_2 t)(-a/2)^t$.

- For $\frac{1}{4}a^2 - b < 0$: $x_t = Ar^t \cos(\theta t + \omega)$,

  where $r = \sqrt{b}$ and $\cos\theta = -\dfrac{a}{2\sqrt{b}}$, $\theta \in [0, \pi]$.

> The solutions of a homogeneous, linear second-order difference equation with constant coefficients a and b. C_1, C_2, and ω are arbitrary constants.

To find a particular solution of

$(*)$ $x_t + ax_{t-1} + bx_{t-2} = c_t$, $b \neq 0$

use the following trial functions and determine the constants by using the method of undetermined coefficients:

10.9

- If $c_t = c$, try $u_t^* = A$.
- If $c_t = ct + d$, try $u_t^* = At + B$.
- If $c_t = t^n$, try $u_t^* = A_0 + A_1 t + \cdots + A_n t^n$.
- If $c_t = c^t$, try $u_t^* = Ac^t$.
- If $c_t = \alpha \sin ct + \beta \cos ct$, try $u_t^* = A \sin ct + B \cos ct$.

> If the function c_t is itself a solution of the homogeneous equation, multiply the trial solution by t. If this new trial function also satisfies the homogeneous equation, multiply the trial function by t again. (See Hildebrand (1968), Sec. 1.8 for the general procedure.)

10.10

$(*)$ $x_t + a_1 x_{t-1} + \cdots + a_n x_{t-n} = b_t$

$(**)$ $x_t + a_1 x_{t-1} + \cdots + a_n x_{t-n} = 0$

> Linear difference equations with constant coefficients.

10.11 $m^n + a_1 m^{n-1} + \cdots + a_{n-1} m + a_n = 0$

> The *characteristic equation* of (10.10). Its roots are called *characteristic roots*.

Suppose the characteristic equation (10.11) has n different roots, $\lambda_1, \ldots, \lambda_n$, and define

$$\theta_r = \frac{\lambda_r}{\displaystyle\prod_{\substack{1 \leq s \leq n \\ s \neq r}} (\lambda_r - \lambda_s)}, \quad r = 1, 2, \ldots, n$$

10.12

Then a special solution of $(10.10)\,(*)$ is given by

$$u_t^* = \sum_{r=1}^{n} \theta_r \sum_{i=0}^{\infty} \lambda_r^i b_{t-i}$$

> The *backward* solution of $(10.10)(*)$, valid if $|\lambda_r| < 1$ for $r = 1, \ldots, n$.

To obtain n linearly independent solutions of (10.10) (**): Find all roots of the characteristic equation (10.11). Then:

- A real root m_i with multiplicity 1 gives rise to a solution m_i^t.

- A real root m_j with multiplicity $p > 1$, gives rise to solutions m_j^t, tm_j^t, ..., $t^{p-1}m_j^t$.

10.13
- A pair of complex roots $m_k = \alpha + i\beta$, $\overline{m}_k = \alpha - i\beta$ with multiplicity 1, gives rise to the solutions $r^t \cos \theta t$, $r^t \sin \theta t$, where $r = \sqrt{\alpha^2 + \beta^2}$, and $\theta \in [0, \pi]$ satisfies $\cos \theta = \alpha/r$, $\sin \theta = \beta/r$.

- A pair of complex roots $m_e = \lambda + i\mu$, $\overline{m}_e = \lambda - i\mu$ with multiplicity $q > 1$ gives rise to the solutions u, v, tu, tv, ..., $t^{q-1}u$, $t^{q-1}v$, with $u = s^t \cos \varphi t$, $v = s^t \sin \varphi t$, where $s = \sqrt{\lambda^2 + \mu^2}$, and $\varphi \in [0, \pi]$ satisfies $\cos \varphi = \lambda/s$, and $\sin \varphi = \mu/s$.

A general method for finding n linearly independent solutions of (10.10) (**).

10.14 The equations in (10.10) are called *(globally asymptotically) stable* if any solution of the homogeneous equation (10.10) (**) approaches 0 as $t \to \infty$.

Definition of stability for a linear equation with constant coefficients.

10.15 The equations in (10.10) are stable if and only if all the roots of the characteristic equation (10.11) have moduli less than 1.

Stability criterion for (10.10).

10.16

$$\begin{vmatrix} 1 & \vdots & a_n \\ \cdots & \vdots & \cdots \\ a_n & \vdots & 1 \end{vmatrix} > 0, \quad \begin{vmatrix} 1 & 0 & \vdots & a_n & a_{n-1} \\ a_1 & 1 & \vdots & 0 & a_n \\ \cdots & & \vdots & & \cdots \\ a_n & 0 & \vdots & 1 & a_1 \\ a_{n-1} & a_n & \vdots & 0 & 1 \end{vmatrix} > 0, \quad \ldots,$$

$$\begin{vmatrix} 1 & 0 & \cdots & 0 & \vdots & a_n & a_{n-1} & \cdots & a_1 \\ a_1 & 1 & \cdots & 0 & \vdots & 0 & a_n & \cdots & a_2 \\ \vdots & \vdots & \ddots & \vdots & \vdots & \vdots & \vdots & \ddots & \vdots \\ a_{n-1} & a_{n-2} & \cdots & 1 & \vdots & 0 & 0 & \cdots & a_n \\ \cdots & & & & \vdots & & & & \cdots \\ a_n & 0 & \cdots & 0 & \vdots & 1 & a_1 & \cdots & a_{n-1} \\ a_{n-1} & a_n & \cdots & 0 & \vdots & 0 & 1 & \cdots & a_{n-2} \\ \vdots & \vdots & \ddots & \vdots & \vdots & \vdots & \vdots & \ddots & \vdots \\ a_1 & a_2 & \cdots & a_n & \vdots & 0 & 0 & \cdots & 1 \end{vmatrix} > 0$$

A necessary and sufficient condition for all the roots of (10.11) to have moduli less than 1. (Schur's theorem.)

10.17 $x_t + a_1 x_{t-1} = b_t$ is stable $\iff |a_1| < 1$

Special case of (10.15) and (10.16).

10.18	$x_t + a_1 x_{t-1} + a_2 x_{t-2} = b_t$ is stable $$\iff \begin{cases} 1 - a_2 > 0 \\ 1 - a_1 + a_2 > 0 \\ 1 + a_1 + a_2 > 0 \end{cases}$$	Special case of (10.15) and (10.16).
10.19	$x_t + a_1 x_{t-1} + a_2 x_{t-2} + a_3 x_{t-3} = b_t$ is stable $$\iff \begin{cases} 3 - a_2 > 0 \\ 1 - a_2 + a_1 a_3 - a_3^2 > 0 \\ 1 + a_2 - |a_1 + a_3| > 0 \end{cases}$$	Special case of (10.15) and (10.16).
10.20	$x_t + a_1 x_{t-1} + a_2 x_{t-2} + a_3 x_{t-3} + a_4 x_{t-4} = b_t$ is stable $\iff$ $$\begin{cases} 1 - a_4 > 0 \\ 3 + 3a_4 - a_2 > 0 \\ 1 + a_2 + a_4 - |a_1 + a_3| > 0 \\ (1 - a_4)^2 (1 + a_4 - a_2) > (a_1 - a_3)(a_1 a_4 - a_3) \end{cases}$$	Special case of (10.15) and (10.16).
10.21	$$x_1(t) = a_{11}(t)x_1(t-1) + \cdots + a_{1n}(t)x_n(t-1) + b_1(t)$$ $$\cdots\cdots\cdots\cdots\cdots\cdots\cdots\cdots\cdots\cdots\cdots\cdots\cdots\cdots\cdots$$ $$x_n(t) = a_{n1}(t)x_1(t-1) + \cdots + a_{nn}(t)x_n(t-1) + b_n(t)$$	*Linear system of difference equations.*
10.22	$\mathbf{x}(t) = \mathbf{A}(t)\mathbf{x}(t-1) + \mathbf{b}(t), \quad t = 1, 2, \ldots$	Matrix form of (10.21). $\mathbf{x}(t)$ and $\mathbf{b}(t)$ are $n \times 1$, $\mathbf{A}(t) = (a_{ij}(t))$ is $n \times n$.
10.23	$\mathbf{x}(t) = \mathbf{A}^t \mathbf{x}(0) + (\mathbf{A}^{t-1} + \mathbf{A}^{t-2} + \cdots + \mathbf{A} + \mathbf{I})\mathbf{b}$	The solution of (10.22) for $\mathbf{A}(t) = \mathbf{A}$, $\mathbf{b}(t) = \mathbf{b}$.
10.24	$\mathbf{x}(t) = \mathbf{A}\mathbf{x}(t-1) \iff \mathbf{x}(t) = \mathbf{A}^t \mathbf{x}(0)$	A special case of (10.23) where $\mathbf{b} = \mathbf{0}$, and with $\mathbf{A}^0 = \mathbf{I}$.
10.25	If $\mathbf{A}$ is an $n \times n$ diagonalizable matrix with eigenvalues $\lambda_1, \lambda_2, \ldots, \lambda_n$, then the solution in (10.24) can be written as $$\mathbf{x}(t) = \mathbf{P} \begin{pmatrix} \lambda_1^t & 0 & \cdots & 0 \\ 0 & \lambda_2^t & \cdots & 0 \\ \vdots & \vdots & \ddots & \vdots \\ 0 & 0 & \cdots & \lambda_n^t \end{pmatrix} \mathbf{P}^{-1}\mathbf{x}(0)$$ where $\mathbf{P}$ is a matrix of corresponding linearly independent eigenvectors of $\mathbf{A}$.	An important result.
10.26	The difference equation (10.22) with $\mathbf{A}(t) = \mathbf{A}$ is called stable if $\mathbf{A}^t \mathbf{x}(0)$ converges to the zero vector for every choice of the vector $\mathbf{x}(0)$.	Definition of *stability* of a linear system.

10.27	The difference equation (10.22) with $\mathbf{A}(t) = \mathbf{A}$ is stable if and only if all the eigenvalues of $\mathbf{A}$ have moduli less than 1.	Characterization of stability of a linear system.

10.28 If all eigenvalues of $\mathbf{A} = (a_{ij})_{n \times n}$ have moduli less than 1, then every solution $\mathbf{x}(t)$ of

$$\mathbf{x}(t) = \mathbf{A}\mathbf{x}(t-1) + \mathbf{b}, \qquad t = 1, 2, \ldots$$

converges to the vector $(\mathbf{I} - \mathbf{A})^{-1}\mathbf{b}$.

The solution of an important equation.

Stability of first-order nonlinear difference equations

10.29 $x_{t+1} = f(x_t), \qquad t = 0, 1, 2, \ldots$

A general first-order difference equation.

10.30 An *equilibrium state* of the difference equation (10.29) is a point x^* such that $f(x^*) = x^*$.

x^* is a fixed point for f. If $x_0 = x^*$, then $x_t = x^*$ for all $t = 0, 1, 2, \ldots$

10.31 An equilibrium state x^* of (10.29) is *locally asymptotically stable* if there exists a $\delta > 0$ such that, if $|x_0 - x^*| < \delta$ then $\lim_{t \to \infty} x_t = x^*$.
   An equilibrium state x^* is *unstable* if there is a $\delta > 0$ such that $|f(x) - x^*| > |x - x^*|$ for every x with $0 < |x - x^*| < \delta$.

A solution of (10.29) that starts sufficiently close to a locally asymptotically stable equilibrium x^* converges to x^*. A solution that starts close to an unstable equilibrium x^* will move away from x^*, at least to begin with.

10.32 If x^* is an equilibrium state for equation (10.29) and f is C^1 in an open interval around x^*, then

- If $|f'(x^*)| < 1$, then x^* is locally asymptotically stable.

- If $|f'(x^*)| > 1$, then x^* is unstable.

A simple criterion for local stability. See figure (10.37) (a).

10.33 A *cycle* or *periodic solution* of $x_{t+1} = f(x_t)$ with minimal period $n > 0$ is a solution such that $x_{t+n} = x_t$ for some t, while $x_{t+k} \neq x_t$ for $k = 1, \ldots, n-1$.

A cycle will repeat itself indefinitely.

10.34 The equation $x_{t+1} = f(x_t)$ admits a cycle of period 2 if and only if there exist points ξ_1 and ξ_2 such that $f(\xi_1) = \xi_2$ and $f(\xi_2) = \xi_1$.

ξ_1 and ξ_2 are fixed points of $F = f \circ f$. See figure (10.37) (b).

68

10.35 A period 2 cycle for $x_{t+1} = f(x_t)$ alternating between ξ_1 and ξ_2 is *locally asymptotically stable* if every solution starting close to ξ_1 (or equivalently ξ_2) converges to the cycle.

The cycle is locally asymptotically stable if ξ_1 and ξ_2 are locally asymptotically stable equilibria of the equation $y_{t+1} = f(f(y_t))$.

10.36 If f is C^1 and $x_{t+1} = f(x_t)$ admits a period 2 cycle ξ_1, ξ_2 then:

- If $|f'(\xi_1)f'(\xi_2)| < 1$, then the cycle is locally asymptotically stable.

- If $|f'(\xi_1)f'(\xi_2)| > 1$, then the cycle is unstable.

An easy consequence of (10.32). The cycle is *unstable* if ξ_1 or ξ_2 (and then both) is an unstable equilibrium of $y_{t+1} = f(f(y_t))$.

10.37

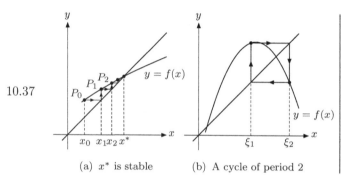

(a) x^* is stable   (b) A cycle of period 2

Illustrations of (10.32) and (10.34). In figure (a), the sequence x_0, x_1, x_2, ... is a solution of (10.29), converging to the equilibrium x^*. The points $P_i = (x_i, x_{i+1})$ are the corresponding points on the graph of f.

References

Most of the formulas and results are found in e.g. Goldberg (1961), Gandolfo (1996), and Hildebrand (1968). For (10.19) and (10.20), see Farebrother (1973). For (10.29)–(10.36), see Sydsæter et al. (2005).

Chapter 11

Differential equations

First-order equations

11.1 $\dot{x}(t) = f(t) \iff x(t) = x(t_0) + \int_{t_0}^{t} f(\tau)\, d\tau$

> A simple differential equation and its solution. $f(t)$ is a given function and $x(t)$ is the unknown function.

11.2 $\dfrac{dx}{dt} = f(t)g(x) \iff \int \dfrac{dx}{g(x)} = \int f(t)\, dt$

Evaluate the integrals. Solve the resulting implicit equation for $x = x(t)$.

> A *separable* differential equation. If $g(a) = 0$, $x(t) \equiv a$ is a solution.

11.3 $\dot{x} = g(x/t)$ and $z = x/t \implies t\dfrac{dz}{dt} = g(z) - z$

> A *projective* differential equation. The substitution $z = x/t$ leads to a separable equation for z.

11.4   The equation $\dot{x} = B(x-a)(x-b)$ has the solutions

$$x \equiv a, \quad x \equiv b, \quad x = a + \frac{b-a}{1 - Ce^{B(b-a)t}}$$

> $a \neq b$. $a = 0$ gives the *logistic* equation. C is a constant.

11.5
- $\dot{x} + ax = b \iff x = Ce^{-at} + \dfrac{b}{a}$
- $\dot{x} + ax = b(t) \iff x = e^{-at}(C + \int b(t)e^{at}\, dt)$

> Linear first-order differential equations with constant coefficient $a \neq 0$. C is a constant.

11.6
$$\dot{x} + a(t)\, x = b(t) \iff$$
$$x = e^{-\int a(t)\, dt}\left(C + \int e^{\int a(t)\, dt} b(t)\, dt\right)$$

> General linear first-order differential equation. $a(t)$ and $b(t)$ are given. C is a constant.

11.7

$$\dot{x} + a(t)\,x = b(t) \iff$$
$$x(t) = x_0 e^{-\int_{t_0}^{t} a(\xi)\,d\xi} + \int_{t_0}^{t} b(\tau) e^{-\int_{\tau}^{t} a(\xi)\,d\xi}\,d\tau$$

Solution of (11.6) with given initial condition $x(t_0) = x_0$.

11.8

$\dot{x} + a(t)x = b(t)x^r$ has the solution
$$x(t) = e^{-A(t)}\left[C + (1-r)\int b(t)e^{(1-r)A(t)}\,dt\right]^{\frac{1}{1-r}}$$
where $A(t) = \int a(t)\,dt$.

Bernoulli's equation and its solution $(r \neq 1)$. C is a constant. (If $r = 1$, the equation is separable.)

11.9

$$\dot{x} = P(t) + Q(t)\,x + R(t)\,x^2$$

Riccati's equation. Not analytically solvable in general. The substitution $x = u + 1/z$ works *if* we know a particular solution $u = u(t)$.

11.10

The differential equation
$$(*) \quad f(t,x) + g(t,x)\,\dot{x} = 0$$
is called *exact* if $f'_x(t,x) = g'_t(t,x)$. The solution $x = x(t)$ is then given implicitly by the equation $\int_{t_0}^{t} f(\tau,x)\,d\tau + \int_{x_0}^{x} g(t_0,\xi)\,d\xi = C$ for some constant C.

An *exact* equation and its solution.

11.11

A function $\beta(t,x)$ is an *integrating factor* for $(*)$ in (11.10) if $\beta(t,x)f(t,x) + \beta(t,x)g(t,x)\dot{x} = 0$ is exact.

- If $(f'_x - g'_t)/g$ is a function of t alone, then $\beta(t) = \exp[\int (f'_x - g'_t)/g\,dt]$ is an integrating factor.

- If $(g'_t - f'_x)/f$ is a function of x alone, then $\beta(x) = \exp[\int (g'_t - f'_x)/f\,dx]$ is an integrating factor.

Results which occasionally can be used to solve equation $(*)$ in (11.10).

11.12

Consider the *initial value problem*
$$(*) \quad \dot{x} = F(t,x), \quad x(t_0) = x_0$$
where $F(t,x)$ and $F'_x(t,x)$ are continuous over the rectangle
$$\Gamma = \{(t,x) : |t - t_0| \leq a, \ |x - x_0| \leq b\}$$
Define
$$M = \max_{(t,x)\in\Gamma} |F(t,x)|, \quad r = \min(a, b/M)$$
Then $(*)$ has a unique solution $x(t)$ on the open interval $(t_0 - r, t_0 + r)$, and $|x(t) - x_0| \leq b$ in this interval.

A *(local) existence and uniqueness theorem.*

Consider the initial value problem

$$\dot{x} = F(t, x), \qquad x(t_0) = x_0$$

Suppose that $F(t, x)$ and $F'_x(t, x)$ are continuous for all (t, x). Suppose too that there exist continuous functions $a(t)$ and $b(t)$ such that

$$(*) \quad |F(t, x)| \leq a(t)|x| + b(t) \quad \text{for all } (t, x)$$

11.13 Given an arbitrary point (t_0, x_0), there exists a unique solution $x(t)$ of the initial value problem, defined on $(-\infty, \infty)$.

Global existence and uniqueness.

If $(*)$ is replaced by the condition

$$xF(t, x) \leq a(t)|x|^2 + b(t) \text{ for all } x \text{ and all } t \geq t_0$$

then the initial value problem has a unique solution defined on $[t_0, \infty)$.

11.14 $\quad \dot{x} = F(x)$

An autonomous first-order differential equation. If $F(a) = 0$, then a is called an *equilibrium*.

11.15

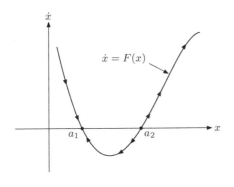

If a solution x starts close to a_1, then $x(t)$ will approach a_1 as t increases. On the other hand, if x starts close to a_2 (but not *at* a_2), then $x(t)$ will move away from a_2 as t increases. a_1 is a *locally stable* equilibrium state for $\dot{x} = F(x)$, whereas a_2 is *unstable*.

11.16
- $F(a) = 0$ and $F'(a) < 0 \Rightarrow a$ is a locally asymptotically stable equilibrium.
- $F(a) = 0$ and $F'(a) > 0 \Rightarrow a$ is an unstable equilibrium.

On stability of equilibrium for (11.14). The precise definitions of stability is given in (11.52).

11.17 If F is a C^1 function, every solution of the autonomous differential equation $\dot{x} = F(x)$ is either constant or strictly monotone on the interval where it is defined.

An interesting result.

Suppose that $x = x(t)$ is a solution of

$$\dot{x} = F(x)$$

11.18

where the function F is continuous. Suppose that $x(t)$ tends to a (finite) limit a as t tends to ∞. Then a must be an equilibrium state for the equation—i.e. $F(a) = 0$.

A convergent solution converges to an equilibrium

Higher order equations

11.19

$$\frac{d^n x}{dt^n} + a_1(t)\frac{d^{n-1}x}{dt^{n-1}} + \cdots + a_{n-1}(t)\frac{dx}{dt} + a_n(t)x = f(t)$$

The general linear nth-order differential equation. When $f(t)$ is not 0, the equation is called inhomogeneous.

11.20

$$\frac{d^n x}{dt^n} + a_1(t)\frac{d^{n-1}x}{dt^{n-1}} + \cdots + a_{n-1}(t)\frac{dx}{dt} + a_n(t)x = 0$$

The homogeneous equation associated with (11.19).

The functions $u_1(t), \ldots, u_m(t)$ are *linearly independent* if the equation

11.21

$$C_1 u_1(t) + \cdots + C_m u_m(t) = 0$$

holds for all t only if the constants $C_1, \ldots, C_m$ are all 0. The functions are *linearly dependent* if they are not linearly independent.

Definition of linear independence and dependence.

If $u_1(t), \ldots, u_n(t)$ are linearly independent solutions of the homogeneous equation (11.20) and $u^*(t)$ is some particular solution of the non-homogeneous equation (11.19), then the general solution of (11.20) is

11.22

$$x(t) = C_1 u_1(t) + \cdots + C_n u_n(t)$$

and the general solution of (11.19) is

$$x(t) = C_1 u_1(t) + \cdots + C_n u_n(t) + u^*(t)$$

where $C_1, \ldots, C_n$ are arbitrary constants.

The structure of the solutions of (11.20) and (11.19). (Note that it is not possible, in general, to find analytic expressions for the required n solutions $u_1(t), \ldots, u_n(t)$ of (11.20).)

Method for finding a particular solution of (11.19) if $u_1, \ldots, u_n$ are n linearly independent solutions of (11.20): Solve the system

$$\dot{C}_1(t)u_1 \quad + \cdots + \quad \dot{C}_n(t)u_n \quad = 0$$
$$\dot{C}_1(t)\dot{u}_1 \quad + \cdots + \quad \dot{C}_n(t)\dot{u}_n \quad = 0$$
$$\cdots\cdots\cdots\cdots\cdots\cdots\cdots\cdots\cdots$$
$$\dot{C}_1(t)u_1^{(n-2)} + \cdots + \dot{C}_n(t)u_n^{(n-2)} = 0$$
$$\dot{C}_1(t)u_1^{(n-1)} + \cdots + \dot{C}_n(t)u_n^{(n-1)} = b(t)$$

for $\dot{C}_1(t), \ldots, \dot{C}_n(t)$. Integrate to find $C_1(t)$, $\ldots$, $C_n(t)$. Then one particular solution of (11.19) is $u^*(t) = C_1(t)u_1 + \cdots + C_n(t)u_n$.

The method of *variation of parameters*, which always makes it possible to find a particular solution of (11.19), provided one knows the general solution of (11.20). Here $u_j^{(i)} = d^i u_j/dt^i$ is the ith derivative of u_j.

11.24

$\ddot{x} + a\dot{x} + bx = 0$ has the general solution:

- If $\frac{1}{4}a^2 - b > 0$: $x = C_1 e^{r_1 t} + C_2 e^{r_2 t}$

  where $r_{1,2} = -\frac{1}{2}a \pm \sqrt{\frac{1}{4}a^2 - b}$.

- If $\frac{1}{4}a^2 - b = 0$: $x = (C_1 + C_2 t)e^{-at/2}$.

- If $\frac{1}{4}a^2 - b < 0$: $x = Ae^{\alpha t}\cos(\beta t + \omega)$,

  where $\alpha = -\frac{1}{2}a$, $\beta = \sqrt{b - \frac{1}{4}a^2}$.

The solution of a homogeneous second-order linear differential equation with constant coefficients a and b. C_1, C_2, A, and ω are constants.

11.25

$\ddot{x} + a\dot{x} + bx = f(t)$, $b \neq 0$, has a particular solution $u^* = u^*(t)$:

- $f(t) = A$: $u^* = A/b$

- $f(t) = At + B$: $u^* = \dfrac{A}{b}t + \dfrac{bB - aA}{b^2}$

- $f(t) = At^2 + Bt + C$:

  $u^* = \frac{A}{b}t^2 + \frac{(bB-2aA)}{b^2}t + \frac{Cb^2 - (2A+aB)b + 2a^2 A}{b^3}$

- $f(t) = pe^{qt}$: $u^* = pe^{qt}/(q^2 + aq + b)$

  (if $q^2 + aq + b \neq 0$).

Particular solutions of $\ddot{x} + a\dot{x} + bx = f(t)$. If $f(t) = pe^{qt}$, $q^2 + aq + b = 0$, and $2q + a \neq 0$, then $u^* = pte^{qt}/(2q + a)$ is a solution. If $f(t) = pe^{qt}$, $q^2 + aq + b = 0$, and $2q + a = 0$, then $u^* = \frac{1}{2}pt^2 e^{qt}$ is a solution.

11.26

$t^2\ddot{x} + at\dot{x} + bx = 0$, $t > 0$, has the general solution:

- If $(a - 1)^2 > 4b$: $x = C_1 t^{r_1} + C_2 t^{r_2}$,

  where $r_{1,2} = -\frac{1}{2}\left[(a-1)\pm\sqrt{(a-1)^2 - 4b}\right]$.

- If $(a - 1)^2 = 4b$: $x = (C_1 + C_2 \ln t)\, t^{(1-a)/2}$.

- If $(a - 1)^2 < 4b$: $x = At^\lambda \cos(\mu \ln t + \omega)$,

  where $\lambda = \frac{1}{2}(1 - a)$, $\mu = \frac{1}{2}\sqrt{4b - (a - 1)^2}$.

The solutions of *Euler's equation* of order 2. C_1, C_2, A, and ω are arbitrary constants.

11.27 $\quad \dfrac{d^n x}{dt^n} + a_1 \dfrac{d^{n-1}x}{dt^{n-1}} + \cdots + a_{n-1}\dfrac{dx}{dt} + a_n x = f(t)$

The general linear differential equation of order n with constant coefficients.

11.28 $\quad \dfrac{d^n x}{dt^n} + a_1 \dfrac{d^{n-1}x}{dt^{n-1}} + \cdots + a_{n-1}\dfrac{dx}{dt} + a_n x = 0$

The *homogeneous equation* associated with (11.27).

11.29 $\quad r^n + a_1 r^{n-1} + \cdots + a_{n-1}r + a_n = 0$

The *characteristic equation* associated with (11.27) and (11.28).

11.30

To obtain n linearly independent solutions of (11.28): Find all roots of (11.29).

- A real root r_i with multiplicity 1 gives rise to a solution $e^{r_i t}$.
- A real root r_j with multiplicity $p > 1$ gives rise to the solutions $e^{r_j t}, te^{r_j t}, \ldots, t^{p-1}e^{r_j t}$.
- A pair of complex roots $r_k = \alpha + i\beta$, $\bar{r}_k = \alpha - i\beta$ with multiplicity 1 gives rise to the solutions $e^{\alpha t}\cos \beta t$ and $e^{\alpha t}\sin \beta t$.
- A pair of complex roots $r_e = \lambda + i\mu$, $\bar{r}_e = \lambda - i\mu$ with multiplicity $q > 1$, gives rise to the solutions u, v, tu, tv, $\ldots$, $t^{q-1}u$, $t^{q-1}v$, where $u = e^{\lambda t}\cos \mu t$ and $v = e^{\lambda t}\sin \mu t$.

General method for finding n linearly independent solutions of (11.28).

11.31 $\quad x = x(t) = C_1 e^{r_1 t} + C_2 e^{r_2 t} + \cdots + C_n e^{r_n t}$

The general solution of (11.28) if the roots $r_1, \ldots, r_n$ of (11.29) are all real and different.

11.32

Equation (11.28) (or (11.27)) is *stable (globally asymptotically stable)* if every solution of (11.28) tends to 0 as $t \to \infty$.

Definition of stability for linear equations with constant coefficients.

11.33

Equation (11.28) is stable $\iff$ all the roots of the characteristic equation (11.29) have negative real parts.

Stability criterion for (11.28).

11.34 $\quad$ (11.28) is stable $\Rightarrow a_i > 0$ for all $i = 1, \ldots, n$

Necessary condition for the stability of (11.28).

$$11.35 \quad \mathbf{A} = \begin{pmatrix} a_1 & a_3 & a_5 & \cdots & 0 & 0 \\ a_0 & a_2 & a_4 & \cdots & 0 & 0 \\ 0 & a_1 & a_3 & \cdots & 0 & 0 \\ \vdots & \vdots & \vdots & \ddots & \vdots & \vdots \\ 0 & 0 & 0 & \cdots & a_{n-1} & 0 \\ 0 & 0 & 0 & \cdots & a_{n-2} & a_n \end{pmatrix}$$

A matrix associated with the coefficients in (11.28) (with $a_0 = 1$). The kth column of this matrix is $\cdots a_{k+1} \, a_k \, a_{k-1} \cdots$, where the element a_k is on the main diagonal. An element a_{k+j} with $k + j$ negative or greater than n, is set to 0.)

$$11.36 \quad (a_1), \quad \begin{pmatrix} a_1 & 0 \\ 1 & a_2 \end{pmatrix}, \quad \begin{pmatrix} a_1 & a_3 & 0 \\ 1 & a_2 & 0 \\ 0 & a_1 & a_3 \end{pmatrix}$$

The matrix $\mathbf{A}$ in (11.35) for $n = 1, 2, 3$, with $a_0 = 1$.

$$11.37 \quad (11.28) \text{ is stable} \iff \begin{cases} \text{all leading principal} \\ \text{minors of } \mathbf{A} \text{ in } (11.35) \\ \text{(with } a_0 = 1) \text{ are pos-} \\ \text{itive.} \end{cases}$$

Routh–Hurwitz's stability conditions.

11.38

- $\dot{x} + a_1 x = f(t)$ is stable $\iff a_1 > 0$
- $\ddot{x} + a_1 \dot{x} + a_2 x = f(t)$ is stable $\iff \begin{cases} a_1 > 0 \\ a_2 > 0 \end{cases}$
- $\dddot{x} + a_1 \ddot{x} + a_2 \dot{x} + a_3 x = f(t)$ is stable
  $\iff a_1 > 0, \; a_3 > 0 \text{ and } a_1 a_2 > a_3$

Special cases of (11.37). (It is easily seen that the conditions are equivalent to requiring that the leading principal minors of the matrices in (11.36) are all positive.)

Systems of differential equations

$$11.39 \quad \left. \begin{array}{l} \dfrac{dx_1}{dt} = f_1(t, x_1, \ldots, x_n) \\ \cdots\cdots\cdots\cdots\cdots \\ \dfrac{dx_n}{dt} = f_n(t, x_1, \ldots, x_n) \end{array} \right\} \iff \dot{\mathbf{x}} = \mathbf{F}(t, \mathbf{x})$$

A *normal* (nonautonomous) system of differential equations. Here $\mathbf{x} = (x_1, \ldots, x_n)$, $\dot{\mathbf{x}} = (\dot{x}_1, \ldots, \dot{x}_n)$, and $\mathbf{F} = (f_1, \cdots, f_n)$.

$$11.40 \quad \begin{array}{l} \dot{x}_1 = a_{11}(t)x_1 + \cdots + a_{1n}(t)x_n + b_1(t) \\ \cdots\cdots\cdots\cdots\cdots\cdots\cdots\cdots\cdots \\ \dot{x}_n = a_{n1}(t)x_1 + \cdots + a_{nn}(t)x_n + b_n(t) \end{array}$$

A linear system of differential equations.

$$11.41 \quad \dot{\mathbf{x}} = \mathbf{A}(t)\mathbf{x} + \mathbf{b}(t), \quad \mathbf{x}(t_0) = \mathbf{x}^0$$

A matrix formulation of (11.40), with an initial condition. $\mathbf{x}, \dot{\mathbf{x}}$, and $\mathbf{b}(t)$ are column vectors and $\mathbf{A}(t) = (a_{ij}(t))_{n \times n}$.

11.42 $\quad \dot{\mathbf{x}} = \mathbf{A}\mathbf{x}, \quad \mathbf{x}(t_0) = \mathbf{x}^0 \iff \mathbf{x} = e^{\mathbf{A}(t-t_0)}\mathbf{x}^0$

The solution of (11.41) for $\mathbf{A}(t) = \mathbf{A}$, $\mathbf{b}(t) = \mathbf{0}$. (For matrix exponentials, see (19.30).)

Let $\mathbf{p}_j(t) = (p_{1j}(t), \ldots, p_{nj}(t))'$, $j = 1, \ldots, n$ be n linearly independent solutions of the homogeneous differential equation $\dot{\mathbf{x}} = \mathbf{A}(t)\mathbf{x}$, with $\mathbf{p}_j(t_0) = \mathbf{e}_j$, $j = 1, \ldots, n$, where $\mathbf{e}_j$ is the jth standard unit vector in $\mathbb{R}^n$. Then the

11.43 *resolvent* of the equation is the matrix

$$\mathbf{P}(t, t_0) = \begin{pmatrix} p_{11}(t) & \cdots & p_{1n}(t) \\ \vdots & \ddots & \vdots \\ p_{n1}(t) & \cdots & p_{nn}(t) \end{pmatrix}$$

The definition of the resolvent of a homogeneous linear differential equation. Note that $\mathbf{P}(t_0, t_0) = \mathbf{I}_n$.

11.44 $\quad \mathbf{x} = \mathbf{P}(t, t_0)\mathbf{x}^0 + \displaystyle\int_{t_0}^{t} \mathbf{P}(t, s)\mathbf{b}(s)\, ds$

The solution of (11.41).

If $\mathbf{P}(t, s)$ is the resolvent of

$$\dot{\mathbf{x}} = \mathbf{A}(t)\mathbf{x}$$

11.45 then $\mathbf{P}(s, t)'$ (the transpose of $\mathbf{P}(s, t)$) is the resolvent of

$$\dot{\mathbf{z}} = -\mathbf{A}(t)'\mathbf{z}$$

A useful fact.

Consider the nth-order differential equation

$$(*) \quad \frac{d^n x}{dt^n} = F\Big(t, x, \frac{dx}{dt}, \ldots, \frac{d^{n-1}x}{dt^{n-1}}\Big)$$

By introducing new variables,

$$y_1 = x, \; y_2 = \frac{dx}{dt}, \; \ldots, \; y_n = \frac{d^{n-1}x}{dt^{n-1}}$$

11.46 one can transform $(*)$ into a normal system:

$$\dot{y}_1 = y_2$$
$$\dot{y}_2 = y_3$$
$$\cdots\cdots$$
$$\dot{y}_{n-1} = y_n$$
$$\dot{y}_n = F(t, y_1, y_2, \ldots, y_n)$$

Any nth-order differential equation can be transformed into a normal system by introducing new unknowns. (A large class of systems of higher order differential equations can be transformed into a normal system by introducing new unknowns in a similar way.)

Consider the initial value problem

(*) $\dot{\mathbf{x}} = \mathbf{F}(t, \mathbf{x}), \quad \mathbf{x}(t_0) = \mathbf{x}^0$

where $\mathbf{F} = (f_1, \ldots, f_n)$ and its first-order partials w.r.t. $x_1, \ldots, x_n$ are continuous over the set

11.47 $\Gamma = \{ (t, \mathbf{x}) : |t - t_0| \le a, \; \|\mathbf{x} - \mathbf{x}^0\| \le b \}$

Define

$$M = \max_{(t,\mathbf{x}) \in \Gamma} \|\mathbf{F}(t, \mathbf{x})\|, \quad r = \min(a, b/M)$$

Then (*) has a unique solution $\mathbf{x}(t)$ on the open interval $(t_0 - r, t_0 + r)$, and $\|\mathbf{x}(t) - \mathbf{x}^0\| \le b$ in this interval.

A (local) existence and uniqueness theorem.

Consider the initial value problem

(1) $\dot{\mathbf{x}} = \mathbf{F}(t, \mathbf{x}), \quad \mathbf{x}(t_0) = \mathbf{x}^0$

where $\mathbf{F} = (f_1, \ldots, f_n)$ and its first-order partials w.r.t. $x_1, \ldots, x_n$ are continuous for all $(t, \mathbf{x})$. Assume, moreover, that there exist continuous functions $a(t)$ and $b(t)$ such that

11.48   (2) $\|\mathbf{F}(t, \mathbf{x})\| \le a(t)\|\mathbf{x}\| + b(t)$ for all $(t, \mathbf{x})$

or

(3) $\mathbf{x} \cdot \mathbf{F}(t, \mathbf{x}) \le a(t)\|\mathbf{x}\|^2 + b(t)$ for all $(t, \mathbf{x})$

Then, given any point $(t_0, \mathbf{x}^0)$, there exists a unique solution $\mathbf{x}(t)$ of (1) defined on $(-\infty, \infty)$.

The inequality (2) *is* satisfied, in particular, if for all $(t, \mathbf{x})$,

(4) $\|\mathbf{F}'_{\mathbf{x}}(t, \mathbf{x})\| \le c(t)$ for a continuous $c(t)$

A global existence and uniqueness theorem. In (4) any matrix norm for $\mathbf{F}'_{\mathbf{x}}(t, \mathbf{x})$ can be used. (For matrix norms, see (19.26).)

Autonomous systems

$\dot{x}_1 = f_1(x_1, \ldots, x_n)$

11.49 $\ldots \ldots \ldots \ldots \ldots \ldots$

$\dot{x}_n = f_n(x_1, \ldots, x_n)$

An autonomous system of first-order differential equations.

11.50 $\mathbf{a} = (a_1, \ldots, a_n)$ is an *equilibrium point* for the system (11.49) if $f_i(\mathbf{a}) = 0$, $i = 1, \ldots, n$.

Definition of an equilibrium point for (11.49).

11.51   If $\mathbf{x}(t) = (x_1(t), \ldots, x_n(t))$ is a solution of the system (11.49) on an interval I, then the set of points $\mathbf{x}(t)$ in $\mathbb{R}^n$ trace out a curve in $\mathbb{R}^n$ called a *trajectory* (or an *orbit*) for the system.

Definition of a trajectory (or an orbit), also called an integral curve.

An equilibrium point **a** for (11.49) is (*locally*) *stable* if all solutions that start close to **a** stay close to **a**: For every $\varepsilon > 0$ there is a $\delta > 0$ such that if $\|\mathbf{x} - \mathbf{a}\| < \delta$, then there exists a solution $\varphi(t)$ of (11.49), defined for $t \geq 0$, with $\varphi(0) = \mathbf{x}$, that satisfies

11.52 $\quad \|\varphi(t) - \mathbf{a}\| < \varepsilon \quad$ for all $t > 0$

Definition of (local) stability and unstability.

If **a** is stable and there exists a $\delta' > 0$ such that

$$\|\mathbf{x} - \mathbf{a}\| < \delta' \implies \lim_{t \to \infty} \|\varphi(t) - \mathbf{a}\| = 0$$

then **a** is (locally) *asymptotically stable*.

If **a** is not stable, it is called *unstable*.

11.53

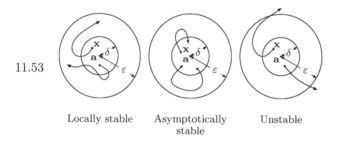

Locally stable   Asymptotically stable   Unstable

Illustrations of stability concepts. The curves with arrows attached are possible trajectories.

11.54   If every solution of (11.49), whatever its initial point, converges to a unique equilibrium point **a**, then **a** is *globally asymptotically stable*.

Global asymptotic stability.

11.55

Locally stable   Globally stable   Unstable

Less technical illustrations of stability concepts.

11.56   Suppose $\mathbf{x}(t)$ is a solution of system (11.49) with $\mathbf{F} = (f_1, \ldots, f_n)$ a C^1 function, and with $\mathbf{x}(t_0 + T) = \mathbf{x}(t_0)$ for some t_0 and some $T > 0$. Then $\mathbf{x}(t + T) = \mathbf{x}(t)$ for *all* t.

If a solution of (11.49) returns to its starting point after a length of time T, then it must be *periodic*, with period T.

11.57 Suppose that a solution $(x(t), y(t))$ of the system

$$\dot{x} = f(x, y), \quad \dot{y} = g(x, y)$$

stays within a compact region of the plane that contains no equilibrium point of the system. Its trajectory must then spiral into a closed curve that is itself the trajectory of a periodic solution of the system.

The Poincaré–Bendixson theorem.

11.58 Let **a** be an equilibrium point for (11.49) and define

$$\mathbf{A} = \begin{pmatrix} \dfrac{\partial f_1(\mathbf{a})}{\partial x_1} & \cdots & \dfrac{\partial f_1(\mathbf{a})}{\partial x_n} \\ \vdots & \ddots & \vdots \\ \dfrac{\partial f_n(\mathbf{a})}{\partial x_1} & \cdots & \dfrac{\partial f_n(\mathbf{a})}{\partial x_n} \end{pmatrix}$$

If all the eigenvalues of **A** have negative real parts, then **a** is (locally) asymptotically stable.

If at least one eigenvalue has a positive real part, then **a** is unstable.

A Liapunov theorem. The equilibrium point **a** is called a *sink* if all the eigenvalues of **A** have negative real parts. (It is called a *source* if all the eigenvalues of **A** have positive real parts.)

11.59 A necessary and sufficient condition for all the eigenvalues of a real $n \times n$ matrix $\mathbf{A} = (a_{ij})$ to have negative real parts is that the following inequalities hold:

- For $n = 2$: $\operatorname{tr}(\mathbf{A}) < 0$ and $|\mathbf{A}| > 0$
- For $n = 3$: $\operatorname{tr}(\mathbf{A}) < 0$, $|\mathbf{A}| < 0$, and

$$\begin{vmatrix} a_{22} + a_{33} & -a_{12} & -a_{13} \\ -a_{21} & a_{11} + a_{33} & -a_{23} \\ -a_{31} & -a_{32} & a_{11} + a_{22} \end{vmatrix} < 0$$

Useful characterizations of stable matrices of orders 2 and 3. (An $n \times n$ matrix is often called *stable* if all its eigenvalues have negative real parts.)

11.60 Let (a, b) be an equilibrium point for the system

$$\dot{x} = f(x, y), \quad \dot{y} = g(x, y)$$

and define

$$\mathbf{A} = \begin{pmatrix} \dfrac{\partial f(a, b)}{\partial x} & \dfrac{\partial f(a, b)}{\partial y} \\ \dfrac{\partial g(a, b)}{\partial x} & \dfrac{\partial g(a, b)}{\partial y} \end{pmatrix}$$

Then, if $\operatorname{tr}(\mathbf{A}) < 0$ and $|\mathbf{A}| > 0$, (a, b) is locally asymptotically stable.

A special case of (11.58). Stability in terms of the signs of the trace and the determinant of **A**, valid if $n = 2$.

11.61 An equilibrium point $\mathbf{a}$ for (11.49) is called *hyperbolic* if the matrix $\mathbf{A}$ in (11.58) has no eigenvalue with real part zero.

Definition of a hyperbolic equilibrium point.

11.62 A hyperbolic equilibrium point for (11.49) is either unstable or asymptotically stable.

An important result.

11.63 Let (a, b) be an equilibrium point for the system
$$\dot{x} = f(x, y), \quad \dot{y} = g(x, y)$$
and define
$$\mathbf{A}(x, y) = \begin{pmatrix} f_1'(x, y) & f_2'(x, y) \\ g_1'(x, y) & g_2'(x, y) \end{pmatrix}$$
Assume that the following three conditions are satisfied:

(a) $\operatorname{tr}(\mathbf{A}(x, y)) = f_1'(x, y) + g_2'(x, y) < 0$
   for all (x, y) in $\mathbb{R}^2$

(b) $|\mathbf{A}(x, y)| = \begin{vmatrix} f_1'(x, y) & f_2'(x, y) \\ g_1'(x, y) & g_2'(x, y) \end{vmatrix} > 0$
   for all (x, y) in $\mathbb{R}^2$

(c) $f_1'(x, y)g_2'(x, y) \neq 0$ for all (x, y) in $\mathbb{R}^2$ or
   $f_2'(x, y)g_1'(x, y) \neq 0$ for all (x, y) in $\mathbb{R}^2$

Then (a, b) is globally asymptotically stable.

Olech's theorem.

11.64 $V(\mathbf{x}) = V(x_1, \ldots, x_n)$ is a Liapunov function for system (11.49) in an open set Ω containing an equilibrium point $\mathbf{a}$ if

- $V(\mathbf{x}) > 0$ for all $\mathbf{x} \neq \mathbf{a}$ in Ω, $V(\mathbf{a}) = 0$, and

- $\dot{V}(\mathbf{x}) = \sum_{i=1}^{n} \frac{\partial V(\mathbf{x})}{\partial x_i} \frac{dx_i}{dt} = \sum_{i=1}^{n} \frac{\partial V(\mathbf{x})}{\partial x_i} f_i(\mathbf{x}) \leq 0$
   for all $\mathbf{x} \neq \mathbf{a}$ in Ω.

Definition of a *Liapunov function*.

11.65 Let $\mathbf{a}$ be an equilibrium point for (11.49) and suppose there exists a Liapunov function $V(\mathbf{x})$ for the system in an open set Ω containing $\mathbf{a}$. Then $\mathbf{a}$ is a stable equilibrium point. If also
$$\dot{V}(\mathbf{x}) < 0 \text{ for all } \mathbf{x} \neq \mathbf{a} \text{ in } \Omega$$
then $\mathbf{a}$ is locally asymptotically stable.

A Liapunov theorem.

The modified Lotka–Volterra model

$$\dot{x} = kx - axy - \varepsilon x^2, \qquad \dot{y} = -hy + bxy - \delta y^2$$

has an asymptotically stable equilibrium

11.66

$$(x_0, y_0) = \left(\frac{ah + k\delta}{ab + \delta\varepsilon}, \frac{bk - h\varepsilon}{ab + \delta\varepsilon} \right)$$

The function $V(x, y) = H(x, y) - H(x_0, y_0)$, where

$$H(x, y) = b(x - x_0 \ln x) + a(y - y_0 \ln y)$$

is a Liapunov function for the system, with $\dot{V}(x, y) < 0$ except at the equilibrium point.

Example of the use of (11.65): x is the number of rabbits, y is the number of foxes. (a, b, h, k, δ, and ε are positive, $bk > h\varepsilon$.) $\varepsilon = \delta = 0$ gives the classical Lotka–Volterra model with $\dot{V} = 0$ everywhere, and integral curves that are closed curves around the equilibrium point.

11.67

Let (a, b) be an equilibrium point for the system

$$\dot{x} = f(x, y), \quad \dot{y} = g(x, y)$$

and define $\mathbf{A}$ as the matrix in (11.60). If $|\mathbf{A}| < 0$, there exist (up to a translation of t) precisely two solutions $(x_1(t), y_1(t))$ and $(x_2(t), y_2(t))$ defined on an interval $[t_0, \infty)$ and converging to (a, b). These solutions converge to (a, b) from opposite directions, and both are tangent to the line through (a, b) parallel to the eigenvector corresponding to the negative eigenvalue. Such an equilibrium is called a *saddle point*.

A *local saddle point theorem*. ($|\mathbf{A}| < 0$ if and only if the eigenvalues of $\mathbf{A}$ are real and of opposite signs.) For a global version of this result, see Seierstad and Sydsæter (1987), Sec. 3.10, Theorem 19.)

Partial differential equations

Method for finding solutions of

$$(*) \quad P(x, y, z)\frac{\partial z}{\partial x} + Q(x, y, z)\frac{\partial z}{\partial y} = R(x, y, z)$$

11.68

- Find the solutions of the system

$$\frac{dy}{dx} = \frac{Q}{P}, \quad \frac{dz}{dx} = \frac{R}{P}$$

where x is the independent variable. If the solutions are given by $y = \varphi_1(x, C_1, C_2)$ and $z = \varphi_2(x, C_1, C_2)$, solve for C_1 and C_2 to obtain $C_1 = u(x, y, z)$ and $C_2 = v(x, y, z)$.

- If Φ is an arbitrary C^1 function of two variables, and at least one of the functions u and v contains z, then $z = z(x, y)$ defined implicitly by the equation

$$\Phi\big(u(x, y, z), v(x, y, z)\big) = 0,$$

is a solution of $(*)$.

The general *quasilinear first-order partial differential equation* and a solution method. The method does not, in general, give *all* the solutions of $(*)$. (See Zachmanoglou and Thoe (1986), Chap. II for more details.)

The following system of partial differential equations

$$\frac{\partial z(\mathbf{x})}{\partial x_1} = f_1(\mathbf{x}, z(\mathbf{x}))$$

$$\frac{\partial z(\mathbf{x})}{\partial x_2} = f_2(\mathbf{x}, z(\mathbf{x}))$$

$$\dots\dots\dots\dots\dots$$

$$\frac{\partial z(\mathbf{x})}{\partial x_n} = f_n(\mathbf{x}, z(\mathbf{x}))$$

in the unknown function $z(\mathbf{x}) = z(x_1, \dots, x_n)$, has a solution if and only if the $n \times n$ matrix of first-order partial derivatives of $f_1, \dots, f_n$ w.r.t. $x_1, \dots, x_n$ is symmetric.

11.69

Frobenius's theorem.
The functions $f_1, \dots, f_n$ are C^1.

References

Braun (1993) is a good reference for ordinary differential equations. For (11.10)–(11.18) see e.g. Sydsæter et al. (2005). For (11.35)–(11.38) see Gandolfo (1996) or Sydsæter et al. (2005). Beavis and Dobbs (1990) have most of the qualitative results and also economic applications. For (11.68) see Sneddon (1957) or Zachmanoglou and Thoe (1986). For (11.69) see Hartman (1982). For economic applications of (11.69) see Mas-Colell, Whinston, and Green (1995).

Chapter 12

Topology in Euclidean space

12.1 $\quad B(\mathbf{a}; r) = \{\, \mathbf{x} : \|\mathbf{x} - \mathbf{a}\| < r \,\} \qquad (r > 0)$

Definition of an *open n-ball* with radius r and center $\mathbf{a}$ in $\mathbb{R}^n$. ($\|\ \|$ is defined in (18.13).)

12.2

- A point $\mathbf{a}$ in $S \subset \mathbb{R}^n$ is an *interior point* of S if there exists an n-ball with center at $\mathbf{a}$, all of whose points belong to S.

- A point $\mathbf{b} \in \mathbb{R}^n$ (not necessarily in S) is a *boundary point* of S if every n-ball with center at $\mathbf{b}$ contains at least one point in S and at least one point not in S.

Definition of interior points and boundary points.

12.3

A set S in $\mathbb{R}^n$ is called

- *open* if all its points are interior points,

- *closed* if $\mathbb{R}^n \setminus S$ is open,

- *bounded* if there exists a number M such that $\|\mathbf{x}\| \leq M$ for all $\mathbf{x}$ in S,

- *compact* if it is closed and bounded.

Important definitions. $\mathbb{R}^n \setminus S$ $= \{\mathbf{x} \in \mathbb{R}^n : \mathbf{x} \notin S\}$.

12.4

A set S in $\mathbb{R}^n$ is closed if and only if it contains all its boundary points. The set $\bar{S}$ consisting of S and all its boundary points is called the *closure* of S.

A useful characterization of closed sets, and a definition of the closure of a set.

12.5

A set S in $\mathbb{R}^n$ is called a *neighborhood* of a point $\mathbf{a}$ in $\mathbb{R}^n$ if $\mathbf{a}$ is an interior point of S.

Definition of a neighborhood.

12.6

A sequence $\{\mathbf{x}_k\}$ in $\mathbb{R}^n$ *converges* to $\mathbf{x}$ if for every $\varepsilon > 0$ there exists an integer N such that $\|\mathbf{x}_k - \mathbf{x}\| < \varepsilon$ for all $k \geq N$.

Convergence of a sequence in $\mathbb{R}^n$. If the sequence does not converge, it *diverges*.

12.7

A sequence $\{\mathbf{x}_k\}$ in $\mathbb{R}^n$ is a *Cauchy sequence* if for every $\varepsilon > 0$ there exists an integer N such that $\|\mathbf{x}_j - \mathbf{x}_k\| < \varepsilon$ for all $j, k \geq N$.

Definition of a Cauchy sequence.

12.8	A sequence $\{\mathbf{x}_k\}$ in $\mathbb{R}^n$ converges if and only if it is a Cauchy sequence.	*Cauchy's convergence criterion.*
12.9	A set S in $\mathbb{R}^n$ is closed if and only if the limit $\mathbf{x} = \lim_k \mathbf{x}_k$ of each convergent sequence $\{\mathbf{x}_k\}$ of points in S also lies in S.	Characterization of a closed set.
12.10	Let $\{\mathbf{x}_k\}$ be a sequence in $\mathbb{R}^n$, and let $k_1 < k_2 < k_3 < \cdots$ be an increasing sequence of integers. Then $\{\mathbf{x}_{k_j}\}_{j=1}^\infty$, is called a *subsequence* of $\{\mathbf{x}_k\}$.	Definition of a subsequence.
12.11	A set S in $\mathbb{R}^n$ is compact if and only if every sequence of points in S has a subsequence that converges to a point in S.	Characterization of a compact set.
12.12	A collection $\mathcal{U}$ of open sets is said to be an *open covering* of the set S if every point of S lies in at least one of the sets from $\mathcal{U}$. The set S has the *finite covering property* if whenever $\mathcal{U}$ is an open covering of S, then a finite subcollection of the sets in $\mathcal{U}$ covers S.	A useful concept.
12.13	A set S in $\mathbb{R}^n$ is compact if and only if it has the finite covering property.	The Heine–Borel theorem.
12.14	$f : M \subset \mathbb{R}^n \to \mathbb{R}$ is *continuous* at $\mathbf{a}$ in M if for every $\varepsilon > 0$ there exists a $\delta > 0$ such that $$|f(\mathbf{x}) - f(\mathbf{a})| < \varepsilon$$ for all $\mathbf{x}$ in M with $\|\mathbf{x} - \mathbf{a}\| < \delta$.	Definition of a continuous function of n variables.
12.15	The function $\mathbf{f} = (f_1, \ldots, f_m) : M \subset \mathbb{R}^n \to \mathbb{R}^m$ is *continuous* at a point $\mathbf{a}$ in M if for every $\varepsilon > 0$ there is a $\delta > 0$ such that $$\|\mathbf{f}(\mathbf{x}) - \mathbf{f}(\mathbf{a})\| < \varepsilon$$ for all $\mathbf{x}$ in M with $\|\mathbf{x} - \mathbf{a}\| < \delta$.	Definition of a continuous vector function of n variables.
12.16	Let $\mathbf{f} = (f_1, \ldots, f_m)$ be a function from $M \subset \mathbb{R}^n$ into $\mathbb{R}^m$, and let $\mathbf{a}$ be a point in M. Then: • $\mathbf{f}$ is continuous at $\mathbf{a}$ if and only if each f_i is continuous at $\mathbf{a}$ according to definition (12.14). • $\mathbf{f}$ is continuous at $\mathbf{a}$ if and only if $\mathbf{f}(\mathbf{x}_k) \to \mathbf{f}(\mathbf{a})$ for every sequence $\{\mathbf{x}_k\}$ in M that converges to $\mathbf{a}$.	Characterizations of a continuous vector function of n variables.

12.17	A function $\mathbf{f} : \mathbb{R}^n \to \mathbb{R}^m$ is continuous at each point $\mathbf{x}$ in $\mathbb{R}^n$ if and only if $\mathbf{f}^{-1}(T)$ is open (closed) for every open (closed) set T in $\mathbb{R}^m$.	Characterization of a continuous vector function from $\mathbb{R}^n$ to $\mathbb{R}^m$.
12.18	If $\mathbf{f}$ is a continuous function of $\mathbb{R}^n$ into $\mathbb{R}^m$ and M is a compact set in $\mathbb{R}^n$, then $\mathbf{f}(M)$ is compact.	Continuous functions map compact sets onto compact sets.
12.19	Given a set A in $\mathbb{R}^n$. The *relative ball* $B^A(\mathbf{a}; r)$ with radius r around $\mathbf{a} \in A$ is defined by the formula $B^A(\mathbf{a}; r) = B(\mathbf{a}; r) \cap A$.	Definition of a relative ball.
12.20	*Relative interior points, relative boundary points, relatively open sets, and relatively closed sets* are defined in the same way as the ordinary versions of these concepts, except that $\mathbb{R}^n$ is replaced by a subset A, and balls by relative balls.	*Relative topology* concepts.
12.21	• $U \subset A$ is relatively open in $A \subset \mathbb{R}^n$ if and only if there exists an open set V in $\mathbb{R}^n$ such that $U = V \cap A$. • $F \subset A$ is relatively closed in $A \subset \mathbb{R}^n$ if and only if there exists a closed set H in $\mathbb{R}^n$ such that $F = H \cap A$.	Characterizations of relatively open and relatively closed subsets of a set $A \subset \mathbb{R}^n$.
12.22	A function $\mathbf{f}$ from $S \subset \mathbb{R}^n$ to $\mathbb{R}^m$ is continuous if and only if either of the following conditions are satisfied: • $\mathbf{f}^{-1}(U)$ is relatively open in S for each open set U in $\mathbb{R}^m$. • $\mathbf{f}^{-1}(T)$ is relatively closed in S for each closed set T in $\mathbb{R}^m$.	A characterization of continuity that applies to functions whose domain is not the whole of $\mathbb{R}^n$.
12.23	A function $\mathbf{f} : M \subset \mathbb{R}^n \to \mathbb{R}^m$ is called *uniformly continuous* on the set $S \subset M$ if for each $\varepsilon > 0$ there exists a $\delta > 0$ (depending on ε but NOT on $\mathbf{x}$ and $\mathbf{y}$) such that $$\|f(\mathbf{x}) - f(\mathbf{y})\| < \varepsilon$$ for all $\mathbf{x}$ and $\mathbf{y}$ in S with $\|\mathbf{x} - \mathbf{y}\| < \delta$.	Definition of uniform continuity of a function from $\mathbb{R}^n$ to $\mathbb{R}^m$.
12.24	If $\mathbf{f} : M \subset \mathbb{R}^n \to \mathbb{R}^m$ is continuous and the set $S \subset M$ is compact, then $\mathbf{f}$ is uniformly continuous on S.	Continuous functions on compact sets are uniformly continuous.

12.25 Let $\{\mathbf{f}_n\}$ be a sequence of functions defined on a set $S \subset \mathbb{R}^n$ and with range in $\mathbb{R}^m$. The sequence $\{\mathbf{f}_n\}$ is said to *converge pointwise* to a function $\mathbf{f}$ on S, if the sequence $\{\mathbf{f}_n(\mathbf{x})\}$ (in $\mathbb{R}^m$) converges to $\mathbf{f}(\mathbf{x})$ for each $\mathbf{x}$ in S.

Definition of (pointwise) convergence of a sequence of functions.

12.26 A sequence $\{\mathbf{f}_n\}$ of functions defined on a set $S \subset \mathbb{R}^n$ and with range in $\mathbb{R}^m$, is said to *converge uniformly* to a function $\mathbf{f}$ on S, if for each $\varepsilon > 0$ there is a natural number $N(\varepsilon)$ (depending on ε but NOT on $\mathbf{x}$) such that

$$\|\mathbf{f}_n(\mathbf{x}) - \mathbf{f}(\mathbf{x})\| < \varepsilon$$

for all $n \geq N(\varepsilon)$ and all $\mathbf{x}$ in S.

Definition of uniform convergence of a sequence of functions.

12.27 A *correspondence* F from a set A to a set B is a rule that maps each x in A to a subset $F(x)$ of B. The *graph* of F is the set

$$\mathrm{graph}(F) = \{(a,b) \in A \times B : b \in F(a)\}$$

Definition of a correspondence and its graph.

12.28 The correspondence $\mathbf{F} : X \subset \mathbb{R}^n \to \mathbb{R}^m$ has *a closed graph* if for every pair of convergent sequences $\{\mathbf{x}_k\}$ in X and $\{\mathbf{y}_k\}$ in $\mathbb{R}^m$ with $\mathbf{y}_k \in \mathbf{F}(\mathbf{x}_k)$ and $\lim_k \mathbf{x}_k = \mathbf{x} \in X$, the limit $\lim_k \mathbf{y}_k$ belongs to $\mathbf{F}(\mathbf{x})$.

Thus $\mathbf{F}$ has a closed graph if and only if $\mathrm{graph}(\mathbf{F})$ is a relatively closed subset of the set $X \times \mathbb{R}^m \subset \mathbb{R}^n \times \mathbb{R}^m$.

Definition of a correspondence with a closed graph.

12.29 The correspondence $\mathbf{F} : X \subset \mathbb{R}^n \to \mathbb{R}^m$ is said to be *lower hemicontinuous* at $\mathbf{x}^0$ if, for each $\mathbf{y}^0$ in $\mathbf{F}(\mathbf{x}^0)$ and each neighborhood U of $\mathbf{y}^0$, there exists a neighborhood N of $\mathbf{x}^0$ such that $\mathbf{F}(\mathbf{x}) \cap U \neq \varnothing$ for all $\mathbf{x}$ in $N \cap X$.

Definition of lower hemicontinuity of a correspondence.

12.30 The correspondence $\mathbf{F} : X \subset \mathbb{R}^n \to \mathbb{R}^m$ is said to be *upper hemicontinuous* at $\mathbf{x}^0$ if for every open set U that contains $\mathbf{F}(\mathbf{x}^0)$, there exists a neighborhood N of $\mathbf{x}^0$ such that $\mathbf{F}(\mathbf{x}) \subset U$ for all x in $N \cap X$.

Definition of upper hemicontinuity of a correspondence.

12.31 Let $\mathbf{F} : X \subset \mathbb{R}^n \to K \subset \mathbb{R}^m$ be a correspondence where K is compact. Suppose that for every $\mathbf{x} \in X$ the set $\mathbf{F}(\mathbf{x})$ is a closed subset of K. Then $\mathbf{F}$ has a closed graph if and only if $\mathbf{F}$ is upper hemicontinuous.

An interesting result.

Infimum and supremum

12.32

- Any non-empty set S of real numbers that is bounded above has a *least upper bound* b^*, i.e. b^* is an upper bound for S and $b^* \leq b$ for every upper bound b of S. b^* is called the *supremum* of S, and we write $b^* = \sup S$.

- Any non-empty set S of real numbers that is bounded below has a *greatest lower bound* a^*, i.e. a^* is a lower bound for S and $a^* \geq a$ for every lower bound a of S. a^* is called the *infimum* of S, and we write $a^* = \inf S$.

The *principle of least upper bound and greatest lower bound* for sets of real numbers. If S is not bounded above, we write $\sup S = \infty$, and if S is not bounded below, we write $\inf S = -\infty$. One usually defines $\sup \varnothing = -\infty$ and $\inf \varnothing = \infty$.

12.33

$$\inf_{\mathbf{x}\in B} f(\mathbf{x}) = \inf\{f(\mathbf{x}) : \mathbf{x} \in B\}$$

$$\sup_{\mathbf{x}\in B} f(\mathbf{x}) = \sup\{f(\mathbf{x}) : \mathbf{x} \in B\}$$

Definition of infimum and supremum of a real valued function defined on a set B in $\mathbb{R}^n$.

12.34

$$\inf_{\mathbf{x}\in B} (f(\mathbf{x}) + g(\mathbf{x})) \geq \inf_{\mathbf{x}\in B} f(\mathbf{x}) + \inf_{\mathbf{x}\in B} g(\mathbf{x})$$

$$\sup_{\mathbf{x}\in B} (f(\mathbf{x}) + g(\mathbf{x})) \leq \sup_{\mathbf{x}\in B} f(\mathbf{x}) + \sup_{\mathbf{x}\in B} g(\mathbf{x})$$

Results about sup and inf.

12.35

$$\inf_{\mathbf{x}\in B} (\lambda f(\mathbf{x})) = \lambda \inf_{\mathbf{x}\in B} f(\mathbf{x}) \quad \text{if } \lambda > 0$$

$$\sup_{\mathbf{x}\in B} (\lambda f(\mathbf{x})) = \lambda \sup_{\mathbf{x}\in B} f(\mathbf{x}) \quad \text{if } \lambda > 0$$

λ is a real number.

12.36

$$\sup_{\mathbf{x}\in B} (-f(\mathbf{x})) = - \inf_{\mathbf{x}\in B} f(\mathbf{x})$$

$$\inf_{\mathbf{x}\in B} (-f(\mathbf{x})) = - \sup_{\mathbf{x}\in B} f(\mathbf{x})$$

12.37

$$\sup_{(\mathbf{x},\mathbf{y})\in A\times B} f(\mathbf{x},\mathbf{y}) = \sup_{\mathbf{x}\in A}(\sup_{\mathbf{y}\in B} f(\mathbf{x},\mathbf{y}))$$

$A \times B = \{(\mathbf{x},\mathbf{y}) : \mathbf{x} \in A \wedge \mathbf{y} \in B\}$

12.38

$$\underline{\lim_{\mathbf{x}\to\mathbf{x}^0}} f(\mathbf{x}) =$$
$$\lim_{r\to 0}(\inf\{f(\mathbf{x}) : 0 < \|\mathbf{x} - \mathbf{x}^0\| < r, \ \mathbf{x} \in M\})$$

$$\overline{\lim_{\mathbf{x}\to\mathbf{x}^0}} f(\mathbf{x}) =$$
$$\lim_{r\to 0}(\sup\{f(\mathbf{x}) : 0 < \|\mathbf{x} - \mathbf{x}^0\| < r, \ \mathbf{x} \in M\})$$

Definition of $\underline{\lim} = \lim\inf$ and $\overline{\lim} = \lim\sup$. f is defined on $M \subset \mathbb{R}^n$ and $\mathbf{x}^0$ is in the closure of $M \setminus \{\mathbf{x}^0\}$.

12.39

$$\underline{\lim}(f + g) \geq \underline{\lim} f + \underline{\lim} g$$

$$\overline{\lim}(f + g) \leq \overline{\lim} f + \overline{\lim} g$$

The inequalities are valid if the right hand sides are defined.

12.40 $\underline{\lim} f \leq \overline{\lim} f$

Results on lim inf and lim sup.

12.41 $\underline{\lim} f = -\overline{\lim}(-f), \quad \overline{\lim} f = -\underline{\lim}(-f)$

Let f be a real valued function defined on the interval $[t_0, \infty)$. Then we define:

12.42
- $\underline{\lim}_{t \to \infty} f(t) = \lim_{t \to \infty} \inf\{f(s) : s \in [t_0, \infty)\}$

- $\overline{\lim}_{t \to \infty} f(t) = \lim_{t \to \infty} \sup\{f(s) : s \in [t_0, \infty)\}$

Definition of $\underline{\lim}_{t \to \infty}$ and $\overline{\lim}_{t \to \infty}$. Formulas (12.39)–(12.41) are still valid.

12.43

- $\underline{\lim}_{t \to \infty} f(t) \geq a \Leftrightarrow \begin{cases} \text{For each } \varepsilon > 0 \text{ there is a} \\ t' \text{ such that } f(t) \geq a - \varepsilon \\ \text{for all } t \geq t'. \end{cases}$

- $\overline{\lim}_{t \to \infty} f(t) \geq a \Leftrightarrow \begin{cases} \text{For each } \varepsilon > 0 \text{ and each} \\ t' \text{ there is a } t \geq t' \text{ such} \\ \text{that } f(t) \geq a - \varepsilon \text{ for all} \\ t \geq t'. \end{cases}$

Basic facts.

References

Bartle (1982), Marsden and Hoffman (1993), and Rudin (1982) are good references for standard topological results. For correspondences and their properties, see Hildenbrand and Kirman (1976) or Hildenbrand (1974).

function is:
 convex $f(a) \leq b$
 concave $f(a) \geq b$

Concave: $f(\alpha x^a + (1-\alpha)x^b) \geq \alpha f(x^a) + (1-\alpha) f(x^b)$
            for all $\alpha \in [0,1]$

If strictly, then just $>$
For convex & strictly convex reverse inequality.

Chapter 13

$\{(x_1, x_2) \in \mathbb{R}^2 \mid x_1 \geq (x_2)^2\}$
convex
strictly convex

$f: \mathbb{R} \rightarrow \mathbb{R}, f(x) = -|x|$
concave
Not strictly

Convexity

$f: \mathbb{R}^2 \rightarrow \mathbb{R}, f(x_1, x_2) = \sqrt{(x_1)^2 + (x_2)^2}$
Lower contour sets are
strictly convex,
strictly quasiconvex

upper
lower

13.1 A set S in $\mathbb{R}^n$ is *convex* if
$$\mathbf{x}, \mathbf{y} \in S \text{ and } \lambda \in [0,1] \Rightarrow \lambda\mathbf{x} + (1-\lambda)\mathbf{y} \in S$$

Definition of a convex set. The empty set is, by definition, convex.

13.2

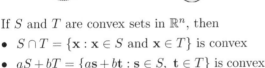

The first set is convex, while the second is not convex.

13.3 If S and T are convex sets in $\mathbb{R}^n$, then
- $S \cap T = \{\mathbf{x} : \mathbf{x} \in S \text{ and } \mathbf{x} \in T\}$ is convex
- $aS + bT = \{a\mathbf{s} + b\mathbf{t} : \mathbf{s} \in S, \mathbf{t} \in T\}$ is convex

Properties of convex sets. (a and b are real numbers.)

13.4 Any vector $\mathbf{x} = \lambda_1\mathbf{x}_1 + \cdots + \lambda_m\mathbf{x}_m$, where $\lambda_i \geq 0$ for $i = 1, \ldots, m$ and $\sum_{i=1}^{m} \lambda_i = 1$, is called a *convex combination* of the vectors $\mathbf{x}_1, \ldots, \mathbf{x}_m$ in $\mathbb{R}^n$.

Definition of a convex combination of vectors.

13.5 $\text{co}(S) = \begin{cases} \text{the set of all convex combinations of} \\ \text{finitely many vectors in } S. \end{cases}$

$\text{co}(S)$ is the *convex hull* of a set S in $\mathbb{R}^n$.

13.6

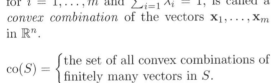

If S is the unshaded set, then $\text{co}(S)$ includes the shaded parts in addition.

13.7 $\text{co}(S)$ is the smallest convex set containing S.

A useful characterization of the convex hull.

13.8 If $S \subset \mathbb{R}^n$ and $\mathbf{x} \in \text{co}(S)$, then $\mathbf{x}$ is a convex combination of at most $n+1$ points in S.

Carathéodory's theorem.

13.9 $\mathbf{z}$ is an *extreme point* of a convex set S if $\mathbf{z} \in S$ and there are no $\mathbf{x}$ and $\mathbf{y}$ in S and λ in $(0,1)$ such that $\mathbf{x} \neq \mathbf{y}$ and $\mathbf{z} = \lambda\mathbf{x} + (1-\lambda)\mathbf{y}$.

Definition of an extreme point.

① $ax_1 + bx_2 \leq c$
lower
not strictly

② $ax_1 + bx_2 = c$
convex
not strictly

③ $\|(x_1, x_2)\| \leq a$ radius
convex
strictly

④ $a \leq ||(x_1, x_2)|| \leq b,\ b > a > 0$
not convex

⑤ $x_1 \leq |x_2|$
not convex

13.10 Let S be a compact, convex set in $\mathbb{R}^n$. Then S is the convex hull of its extreme points.

Krein–Milman's theorem.

13.11 Let S and T be two disjoint non-empty convex sets in $\mathbb{R}^n$. Then S and T can be separated by a hyperplane, i.e. there exists a non-zero vector $\mathbf{a}$ such that

$$\mathbf{a} \cdot \mathbf{x} \leq \mathbf{a} \cdot \mathbf{y} \quad \text{for all } \mathbf{x} \text{ in } S \text{ and all } \mathbf{y} \text{ in } T$$

Minkowski's separation theorem. A hyperplane $\{\mathbf{x} : \mathbf{a} \cdot \mathbf{x} = A\}$, with $\mathbf{a} \cdot \mathbf{x} \leq A \leq \mathbf{a} \cdot \mathbf{y}$ for all $\mathbf{x}$ in S and all $\mathbf{y}$ in T, is called separating.

13.12

In the first figure S and T are (strictly) separated by H. In the second, S and T cannot be separated by a hyperplane.

13.13 Let S be a convex set in $\mathbb{R}^n$ with interior points and let T be a convex set in $\mathbb{R}^n$ such that no point in $S \cap T$ (if there are any) is an interior point of S. Then S and T can be separated by a hyperplane, i.e. there exists a vector $\mathbf{a} \neq \mathbf{0}$ such that

$$\mathbf{a} \cdot \mathbf{x} \leq \mathbf{a} \cdot \mathbf{y} \text{ for all } \mathbf{x} \text{ in } S \text{ and all } \mathbf{y} \text{ in } T.$$

A general separation theorem in $\mathbb{R}^n$.

Concave and convex functions

13.14 $f(\mathbf{x}) = f(x_1, \ldots, x_n)$ defined on a convex set S in $\mathbb{R}^n$ is *concave* on S if

$$f(\lambda \mathbf{x} + (1 - \lambda)\mathbf{x}^0) \geq \lambda f(\mathbf{x}) + (1 - \lambda)f(\mathbf{x}^0)$$

for all $\mathbf{x}, \mathbf{x}^0$ in S and all λ in $(0, 1)$.

To define a *convex* function, reverse the inequality. Equivalently, f is convex if and only if $-f$ is concave.

13.15

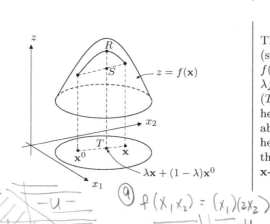

The function $f(\mathbf{x})$ is (strictly) concave. $TR = f(\lambda\mathbf{x} + (1-\lambda)\mathbf{x}^0) \geq TS = \lambda f(\mathbf{x}) + (1 - \lambda)f(\mathbf{x}^0)$. ($TR$ and TS are the heights of R and S above the $\mathbf{x}$-plane. The heights are negative if the points are below the $\mathbf{x}$-plane.)

⑧ $f(x_1, x_2) = x_1 + 2x_2$
Concave
convex
Q–concave
Q–convex

⑨ $f(x_1, x_2) = (x_1)(2x_2)$
$u = $ strictly convex function n
strictly quasi-concave

⑥ $f(x) = ax^2$, $A > 0$
convex
strictly
not concave

⑨ $f(x) = a\sqrt{x}$, $a > 0$
concave
strictly

91

13.16 $f(\mathbf{x})$ is *strictly concave* if $f(\mathbf{x})$ is concave and the inequality $\geq$ in (13.14) is strict for $\mathbf{x} \neq \mathbf{x}^0$.

Definition of a strictly concave function. For strict convexity, reverse the inequality.

13.17 If $f(\mathbf{x})$, defined on the convex set S in $\mathbb{R}^n$, is concave (convex), then $f(\mathbf{x})$ is continuous at each interior point of S.

On the continuity of concave and convex functions.

13.18
- If $f(\mathbf{x})$ and $g(\mathbf{x})$ are concave (convex) and a and b are nonnegative numbers, then $af(\mathbf{x}) + bg(\mathbf{x})$ is concave (convex).

- If $f(\mathbf{x})$ is concave and $F(u)$ is concave and increasing, then $U(\mathbf{x}) = F(f(\mathbf{x}))$ is concave.

- If $f(\mathbf{x}) = \mathbf{a} \cdot \mathbf{x} + b$ and $F(u)$ is concave, then $U(\mathbf{x}) = F(f(\mathbf{x}))$ is concave.

- If $f(\mathbf{x})$ is convex and $F(u)$ is convex and increasing, then $U(\mathbf{x}) = F(f(\mathbf{x}))$ is convex.

- If $f(\mathbf{x}) = \mathbf{a} \cdot \mathbf{x} + b$ and $F(u)$ is convex, then $U(\mathbf{x}) = F(f(\mathbf{x}))$ is convex.

Properties of concave and convex functions.

13.19 A C^1 function $f(\mathbf{x})$ is concave on an open, convex set S of $\mathbb{R}^n$ if and only if

$$f(\mathbf{x}) - f(\mathbf{x}^0) \leq \sum_{i=1}^{n} \frac{\partial f(\mathbf{x}^0)}{\partial x_i}(x_i - x_i^0)$$

or, equivalently,

$$f(\mathbf{x}) - f(\mathbf{x}^0) \leq \nabla f(\mathbf{x}^0) \cdot (\mathbf{x} - \mathbf{x}^0)$$

for all $\mathbf{x}$ and $\mathbf{x}_0$ in S.

Concavity for C^1 functions. For convexity, reverse the inequalities.

13.20 A C^1 function $f(\mathbf{x})$ is strictly concave on an open, convex set S in $\mathbb{R}^n$ if and only if the inequalities in (13.19) are strict for $\mathbf{x} \neq \mathbf{x}^0$.

Strict concavity for C^1 functions. For strict convexity, reverse the inequalities.

13.21 A C^1 function $f(x)$ is concave on an open interval I if and only if

$$f(x) - f(x^0) \leq f'(x^0)(x - x^0)$$

for all x and x^0 in I.

One-variable version of (13.19).

②
$f(x_1, x_2) = 2$
all convex
$=$ strictly
etc. Q-convex

⑪ $f(x_1, x_2) = (x_1)^2 + 2(x_2)^2$
L = 4
L = Strictly convex
Function - Q-convex
strictly Q-convex

13.22

Geometric interpretation of (13.21). The C^1 function f is concave if and only if the graph of f is below the tangent at any point. (In the figure, f is actually strictly concave.)

A C^1 function $f(x, y)$ is concave on an open, convex set S in the (x, y)-plane if and only if

13.23

$$f(x, y) - f(x^0, y^0)$$
$$\leq f_1'(x^0, y^0)(x - x^0) + f_2'(x^0, y^0)(y - y^0)$$

for all (x, y), (x^0, y^0) in S.

Two-variable version of (13.19).

13.24
$$\mathbf{f}''(\mathbf{x}) = \begin{pmatrix} f_{11}''(\mathbf{x}) & f_{12}''(\mathbf{x}) & \cdots & f_{1n}''(\mathbf{x}) \\ f_{21}''(\mathbf{x}) & f_{22}''(\mathbf{x}) & \cdots & f_{2n}''(\mathbf{x}) \\ \vdots & \vdots & \ddots & \vdots \\ f_{n1}''(\mathbf{x}) & f_{n2}''(\mathbf{x}) & \cdots & f_{nn}''(\mathbf{x}) \end{pmatrix}$$

The *Hessian matrix* of f at $\mathbf{x}$. If f is C^2, then the Hessian is symmetric.

13.25

The principal minors $\Delta_r(\mathbf{x})$ of order r in the Hessian matrix $\mathbf{f}''(\mathbf{x})$ are the determinants of the sub-matrices obtained by deleting $n - r$ arbitrary rows and then deleting the $n-r$ columns having the same numbers.

The principal minors of the Hessian. (See also (20.15).)

13.26

A C^2 function $f(\mathbf{x})$ is concave on an open, convex set S in $\mathbb{R}^n$ if and only if for all $\mathbf{x}$ in S and for all Δ_r,

$$(-1)^r \Delta_r(\mathbf{x}) \geq 0 \quad \text{for} \quad r = 1, \ldots, n$$

Concavity for C^2 functions.

13.27

A C^2 function $f(\mathbf{x})$ is convex on an open, convex set S in $\mathbb{R}^n$ if and only if for all $\mathbf{x}$ in S and for all Δ_r,

$$\Delta_r(\mathbf{x}) \geq 0 \quad \text{for} \quad r = 1, \ldots, n$$

Convexity for C^2 functions.

13.28
$$D_r(\mathbf{x}) = \begin{vmatrix} f_{11}''(\mathbf{x}) & f_{12}''(\mathbf{x}) & \cdots & f_{1r}''(\mathbf{x}) \\ f_{21}''(\mathbf{x}) & f_{22}''(\mathbf{x}) & \cdots & f_{2r}''(\mathbf{x}) \\ \vdots & \vdots & \ddots & \vdots \\ f_{r1}''(\mathbf{x}) & f_{r2}''(\mathbf{x}) & \cdots & f_{rr}''(\mathbf{x}) \end{vmatrix}$$

The leading principal minors of the Hessian matrix of f at $\mathbf{x}$, where $r = 1, 2, \ldots, n$.

A C^2 function $f(\mathbf{x})$ is strictly concave on an open, convex set S in $\mathbb{R}^n$ if for all $\mathbf{x} \in S$,

$$(-1)^r D_r(\mathbf{x}) > 0 \quad \text{for} \quad r = 1, \ldots, n$$

Sufficient (but NOT necessary) conditions for strict concavity for C^2 functions.

13.29

A C^2 function $f(\mathbf{x})$ is strictly convex on an open, convex set S in $\mathbb{R}^n$ if for all $\mathbf{x} \in S$,

$$D_r(\mathbf{x}) > 0 \quad \text{for} \quad r = 1, \ldots, n$$

Sufficient (but NOT necessary) conditions for strict convexity for C^2 functions.

13.30

Suppose $f(x)$ is a C^2 function on an open interval I. Then:

- $f(x)$ is concave on $I \Leftrightarrow f''(x) \leq 0$ for all x in I
- $f(x)$ is convex on $I \Leftrightarrow f''(x) \geq 0$ for all x in I
- $f''(x) < 0$ for all x in $I \Rightarrow f(x)$ is strictly concave on I
- $f''(x) > 0$ for all x in $I \Rightarrow f(x)$ is strictly convex on I

One-variable versions of (13.26), (13.27), (13.29), and (13.30). The implication arrows CANNOT be replaced by equivalence arrows. ($f(x) = -x^4$ is strictly concave, but $f''(0) = 0$. $f(x) = x^4$ is strictly convex, but $f''(0) = 0$.)

13.31

A C^2 function $f(x, y)$ is concave on an open, convex set S in the (x, y)-plane if and only if

$$f''_{11}(x, y) \leq 0, \quad f''_{22}(x, y) \leq 0 \quad \text{and}$$
$$f''_{11}(x, y)f''_{22}(x, y) - (f''_{12}(x, y))^2 \geq 0$$

for all (x, y) in S.

Two-variable version of (13.26). For convexity of C^2 functions, reverse the first two inequalities.

13.32

A C^2 function $f(x, y)$ is strictly concave on an open, convex set S in the (x, y)-plane if (but NOT only if)

$$f''_{11}(x, y) < 0 \quad \text{and}$$
$$f''_{11}(x, y)f''_{22}(x, y) - (f''_{12}(x, y))^2 > 0$$

for all (x, y) in S.

Two-variable version of (13.29). (Note that the two inequalities imply $f''_{22}(x, y) < 0$.) For strict convexity, reverse the first inequality.

13.33

Quasiconcave and quasiconvex functions

$f(\mathbf{x})$ is *quasiconcave* on a convex set $S \subset \mathbb{R}^n$ if the *(upper) level set*

$$P_a = \{\mathbf{x} \in S : f(\mathbf{x}) \geq a\}$$

is convex for each real number a.

Definition of a quasiconcave function. (Upper level sets are also called upper contour sets.)

13.34

Quasiconvex — not anything (handwritten annotations)

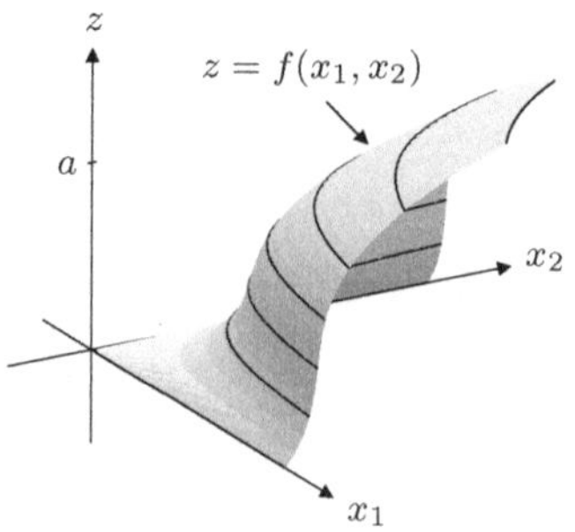

A typical example of a quasiconcave function of two variables, $z = f(x_1, x_2)$.

13.35

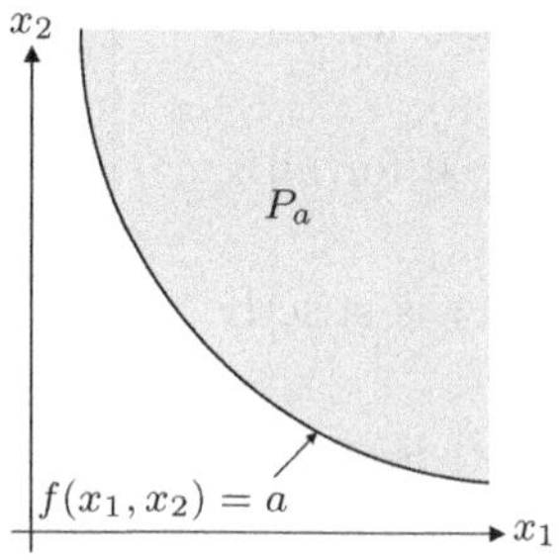

An (upper) level set for the function in (13.35), $P_a = \{(x_1, x_2) \in S : f(x_1, x_2) \geq a\}$.

13.36

13.37

$f(\mathbf{x})$ is quasiconcave on an open, convex set S in $\mathbb{R}^n$ if and only if

$$f(\mathbf{x}) \geq f(\mathbf{x}^0) \Rightarrow f(\lambda\mathbf{x} + (1-\lambda)\mathbf{x}^0) \geq f(\mathbf{x}^0)$$

for all $\mathbf{x}$, $\mathbf{x}^0$ in S and all λ in $[0,1]$.

Characterization of quasiconcavity.

13.38

$f(\mathbf{x})$ is *strictly quasiconcave* on an open, convex set S in $\mathbb{R}^n$ if

$$f(\mathbf{x}) \geq f(\mathbf{x}^0) \Rightarrow f(\lambda\mathbf{x} + (1-\lambda)\mathbf{x}^0) > f(\mathbf{x}^0)$$

for all $\mathbf{x} \neq \mathbf{x}^0$ i S and all λ in $(0,1)$.

The (most common) definition of strict quasiconcavity.

13.39

$f(\mathbf{x})$ is *(strictly) quasiconvex* on $S \subset \mathbb{R}^n$ if $-f(\mathbf{x})$ is (strictly) quasiconcave.

Definition of a (strictly) quasiconvex function.

13.40

If $f_1, \ldots, f_m$ are concave functions defined on a convex set S in $\mathbb{R}^n$ and g is defined for each $\mathbf{x}$ in S by

$$g(\mathbf{x}) = F(f_1(\mathbf{x}), \ldots, f_m(\mathbf{x}))$$

with $F(u_1, \ldots, u_m)$ quasiconcave and increasing in each variable, then g is quasiconcave.

A useful result.

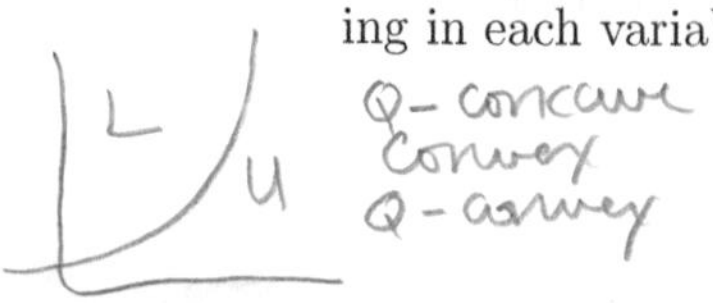

(handwritten notes at bottom)

Q-concave
Convex
Q-convex

concave
Q concave
not convex
Q-convex

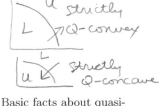

strictly
Q-convex

strictly
Q-concave

Basic facts about quasi-
concave and quasicon-
vex functions. (Exam-
ple of (4): $f(x) = x^3$
and $g(x) = -x$ are
both quasiconcave, but
$f(x) + g(x) = x^3 - x$
is not.) For a proof of
(7), see Sydsæter et al.
(2005).

(1) $f(\mathbf{x})$ concave $\Rightarrow$ $f(\mathbf{x})$ quasiconcave.

(2) $f(\mathbf{x})$ convex $\Rightarrow$ $f(\mathbf{x})$ quasiconvex.

(3) Any increasing or decreasing function of one variable is quasiconcave and quasiconvex.

(4) A sum of quasiconcave (quasiconvex) functions is not necessarily quasiconcave (quasiconvex).

13.41

(5) If $f(\mathbf{x})$ is quasiconcave (quasiconvex) and F is increasing, then $F(f(\mathbf{x}))$ is quasiconcave (quasiconvex).

(6) If $f(\mathbf{x})$ is quasiconcave (quasiconvex) and F is decreasing, then $F(f(\mathbf{x}))$ is quasiconvex (quasiconcave).

(7) Let $f(\mathbf{x})$ be a function defined on a convex cone K in $\mathbb{R}^n$. Suppose that f is quasiconcave and homogeneous of degree q, where $0 < q \leq 1$, that $f(\mathbf{0}) = 0$, and that $f(\mathbf{x}) > 0$ for all $\mathbf{x} \neq \mathbf{0}$ in K. Then f is concave.

13.42

A C^1 function $f(\mathbf{x})$ is quasiconcave on an open, convex set S in $\mathbb{R}^n$ if and only if

$$f(\mathbf{x}) \geq f(\mathbf{x}^0) \Rightarrow \nabla f(\mathbf{x}^0) \cdot (\mathbf{x} - \mathbf{x}^0) \geq 0$$

for all $\mathbf{x}$ and $\mathbf{x}^0$ in S.

Quasiconcavity for
C^1 functions.

13.43

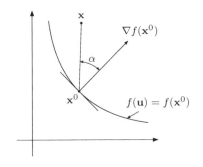

A geometric interpre-
tation of (13.42). Here
$\nabla f(\mathbf{x}^0) \cdot (\mathbf{x} - \mathbf{x}^0) \geq 0$
means that the angle α
is acute, i.e. less than
$90°$.

13.44 $$B_r(\mathbf{x}) = \begin{vmatrix} 0 & f_1'(\mathbf{x}) & \cdots & f_r'(\mathbf{x}) \\ f_1'(\mathbf{x}) & f_{11}''(\mathbf{x}) & \cdots & f_{1r}''(\mathbf{x}) \\ \vdots & \vdots & \ddots & \vdots \\ f_r'(\mathbf{x}) & f_{r1}''(\mathbf{x}) & \cdots & f_{rr}''(\mathbf{x}) \end{vmatrix}$$

A *bordered Hessian* associated with f at $\mathbf{x}$.

13.45

If $f(\mathbf{x})$ is quasiconcave on an open, convex set S in $\mathbb{R}^n$, then

$$(-1)^r B_r(\mathbf{x}) \geq 0 \text{ for } r = 1, \ldots, n$$

for all $\mathbf{x} \in S$.

Necessary conditions
for quasiconcavity of C^2
functions.

13.46	If $(-1)^r B_r(\mathbf{x}) > 0$ for $r = 1, \ldots, n$ for all $\mathbf{x}$ in an open, convex set S in $\mathbb{R}^n$, then $f(\mathbf{x})$ is quasiconcave in S.	Sufficient conditions for quasiconcavity of C^2 functions.
13.47	If $f(\mathbf{x})$ is quasiconvex on an open, convex set S in $\mathbb{R}^n$, then $$B_r(\mathbf{x}) \leq 0 \text{ for } r = 1, \ldots, n$$ and for all $\mathbf{x}$ in S.	Necessary conditions for quasiconvexity of C^2 functions.
13.48	If $B_r(\mathbf{x}) < 0$ for $r = 1, \ldots, n$ and for all $\mathbf{x}$ in an open, convex set S in $\mathbb{R}^n$, then $f(\mathbf{x})$ is quasiconvex in S.	Sufficient conditions for quasiconvexity of C^2 functions.

Pseudoconcave and pseudoconvex functions

13.49	A C^1 function $f(\mathbf{x})$ defined on a convex set S in $\mathbb{R}^n$ is *pseudoconcave* at the point $\mathbf{x}^0$ in S if $$(*) \quad f(\mathbf{x}) > f(\mathbf{x}^0) \Rightarrow \nabla f(\mathbf{x}^0) \cdot (\mathbf{x} - \mathbf{x}^0) > 0$$ for all $\mathbf{x}$ in S. $f(\mathbf{x})$ is *pseudoconcave over S* if $(*)$ holds for all $\mathbf{x}$ and $\mathbf{x}^0$ in S.	To define pseudoconvex functions, reverse the second inequality in $(*)$. (Compare with the characterization of quasiconcavity in (13.42).)
13.50	Let $f(\mathbf{x})$ be a C^1 function defined on a convex set S in $\mathbb{R}^n$. Then: • If f is pseudoconcave on S, then f is quasiconcave on S. • If S is open and if $\nabla f(\mathbf{x}) \neq \mathbf{0}$ for all $\mathbf{x}$ in S, then f is pseudoconcave on S if and only if f is quasiconcave on S.	Important relationships between pseudoconcave and quasiconcave functions.
13.51	Let S be an open, convex set in $\mathbb{R}^n$, and let $f : S \to \mathbb{R}$ be a pseudoconcave function. If $\mathbf{x}^0 \in S$ has the property that $$\nabla f(\mathbf{x}^0) \cdot (\mathbf{x} - \mathbf{x}^0) \leq 0 \text{ for all } \mathbf{x} \text{ in } S$$ (which *is* the case if $\nabla f(\mathbf{x}^0) = \mathbf{0}$), then $\mathbf{x}^0$ is a global maximum point for f in S.	Shows one reason for introducing the concept of pseudoconcavity.

References

For concave/convex and quasiconcave/quasiconvex functions, see e.g. Simon and Blume (1994) or Sydsæter et al. (2005). For pseudoconcave and pseudoconvex functions, see e.g. Simon and Blume (1994), and their references. For special results on convex sets, see Nikaido (1968) and Takayama (1985). A standard reference for convexity theory is Rockafellar (1970).

Chapter 14

Classical optimization

14.1
$f(\mathbf{x}) = f(x_1, \ldots, x_n)$ has a *maximum (minimum)* at $\mathbf{x}^* = (x_1^*, \ldots, x_n^*) \in S$ if

$f(\mathbf{x}^*) - f(\mathbf{x}) \geq 0 \ (\leq 0)$ for all $\mathbf{x}$ in S

$\mathbf{x}^*$ is called a *maximum (minimum) point* and $f(\mathbf{x}^*)$ is called a *maximum (minimum) value*.

Definition of (global) maximum (minimum) of a function of n variables. As collective names, we use *optimal* points and values, or *extreme* points and values.

14.2
$\mathbf{x}^*$ maximizes $f(\mathbf{x})$ over S if and only if $\mathbf{x}^*$ minimizes $-f(\mathbf{x})$ over S.

Used to convert minimization problems to maximization problems.

14.3
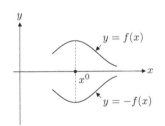

Illustration of (14.2). x^* maximizes $f(x)$ if and only if x^* minimizes $-f(x)$

14.4
Suppose $f(\mathbf{x})$ is defined on $S \subset \mathbb{R}^n$ and that $F(u)$ is strictly increasing on the range of f. Then $\mathbf{x}^*$ maximizes (minimizes) $f(\mathbf{x})$ on S if and only if $\mathbf{x}^*$ maximizes (minimizes) $F(f(\mathbf{x}))$ on S.

An important fact.

14.5
If $f : S \to \mathbb{R}$ is continuous on a closed, bounded set S in $\mathbb{R}^n$, then there exist maximum and minimum points for f in S.

The *extreme value theorem* (or *Weierstrass's theorem*).

14.6
$\mathbf{x}^* = (x_1^*, \ldots, x_n^*)$ is a *stationary* point of $f(\mathbf{x})$ if

$f_1'(\mathbf{x}^*) = 0, \ f_2'(\mathbf{x}^*) = 0, \ \ldots, \ f_n'(\mathbf{x}^*) = 0$

Definition of stationary points for a differentiable function of n variables.

14.7
Let $f(\mathbf{x})$ be concave (convex) and defined on a convex set S in $\mathbb{R}^n$, and let $\mathbf{x}^*$ be an interior point of S. Then $\mathbf{x}^*$ maximizes (minimizes) $f(\mathbf{x})$ on S, if and only if $\mathbf{x}^*$ is a stationary point.

Maximum (minimum) of a concave (convex) function.

14.8

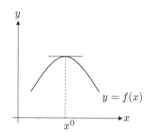

One-variable illustration of (14.7). f is concave, $f'(x^*) = 0$, and x^* is a maximum point.

14.9

If $f(\mathbf{x})$ has a maximum or minimum in $S \subset \mathbb{R}^n$, then the maximum/minimum points are found among the following points:

- interior points of S that are stationary
- extreme points of f at the boundary of S
- points in S where f is not differentiable

Where to find (global) maximum or minimum points.

14.10

$f(\mathbf{x})$ has a *local* maximum (minimum) at $\mathbf{x}^*$ if

$(*) \quad f(\mathbf{x}^*) - f(\mathbf{x}) \geq 0 \ (\leq 0)$

for all $\mathbf{x}$ in S sufficiently close to $\mathbf{x}^*$. More precisely, there exists an n-ball $B(\mathbf{x}^*; r)$ such that $(*)$ holds for all $\mathbf{x}$ in $S \cap B(\mathbf{x}^*; r)$.

Definition of local (or *relative*) maximum (minimum) points of a function of n variables. A collective name is *local extreme points*.

14.11

If $f(\mathbf{x}) = f(x_1, \ldots, x_n)$ has a local maximum (minimum) at an interior point $\mathbf{x}^*$ of S, then $\mathbf{x}^*$ is a stationary point of f.

The *first-order conditions* for differentiable functions.

14.12

A stationary point $\mathbf{x}^*$ of $f(\mathbf{x}) = f(x_1, \ldots, x_n)$ is called a *saddle point* if it is neither a local maximum point nor a local minimum point, i.e. if every n-ball $B(\mathbf{x}^*; r)$ contains points $\mathbf{x}$ such that $f(\mathbf{x}) < f(\mathbf{x}^*)$ and other points $\mathbf{z}$ such that $f(\mathbf{z}) > f(\mathbf{x}^*)$.

Definition of a saddle point.

14.13

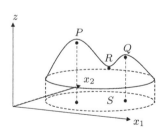

The points P, Q, and R are all stationary points. P is a maximum point, Q is a local maximum point, whereas R is a saddle point.

Special results for one-variable functions

14.14

If $f(x)$ is differentiable in an interval I, then

- $f'(x) > 0 \implies f(x)$ is strictly increasing
- $f'(x) \geq 0 \iff f(x)$ is increasing
- $f'(x) = 0 \iff f(x)$ is constant
- $f'(x) \leq 0 \iff f(x)$ is decreasing
- $f'(x) < 0 \implies f(x)$ is strictly decreasing

Important facts. The implication arrows cannot be reversed. ($f(x) = x^3$ is strictly increasing, but $f'(0) = 0$. $g(x) = -x^3$ is strictly decreasing, but $g'(0) = 0$.)

14.15

- If $f'(x) \geq 0$ for $x \leq c$ and $f'(x) \leq 0$ for $x \geq c$, then $x = c$ is a maximum point for f.
- If $f'(x) \leq 0$ for $x \leq c$, and $f'(x) \geq 0$ for $x \geq c$, then $x = c$ is a minimum point for f.

A first-derivative test for (global) max/min. (Often ignored in elementary mathematics for economics texts.)

14.16

One-variable illustrations of (14.15). c is a maximum point. d is a minimum point.

14.17

c is an *inflection point* for $f(x)$ if $f''(x)$ changes sign at c.

Definition of an inflection point for a function of one variable.

14.18

An unorthodox illustration of an inflection point. Point P, where the slope is steepest, is an inflection point.

14.19

Let f be a function with a continuous second derivative in an interval I, and suppose that c is an interior point of I. Then:

- c is an inflection point for $f \implies f''(c) = 0$
- $f''(c) = 0$ and f'' changes sign at c
  $\implies c$ is an inflection point for f

Test for inflection points.

Second-order conditions

If $f(\mathbf{x}) = f(x_1, \ldots, x_n)$ has a local maximum (minimum) at $\mathbf{x}^*$, then

14.20
$$\sum_{i=1}^{n} \sum_{j=1}^{n} f_{ij}''(\mathbf{x}^*) h_i h_j \leq 0 \ (\geq 0)$$

for all choices of $h_1, \ldots, h_n$.

A necessary (second-order) condition for local maximum (minimum).

If $\mathbf{x}^* = (x_1^*, \ldots, x_n^*)$ is a stationary point of $f(x_1, \ldots, x_n)$, and if $D_k(\mathbf{x}^*)$ is the following determinant,

$$D_k(\mathbf{x}^*) = \begin{vmatrix} f_{11}''(\mathbf{x}^*) & f_{12}''(\mathbf{x}^*) & \cdots & f_{1k}''(\mathbf{x}^*) \\ f_{21}''(\mathbf{x}^*) & f_{22}''(\mathbf{x}^*) & \cdots & f_{2k}''(\mathbf{x}^*) \\ \vdots & \vdots & \ddots & \vdots \\ f_{k1}''(\mathbf{x}^*) & f_{k2}''(\mathbf{x}^*) & \cdots & f_{kk}''(\mathbf{x}^*) \end{vmatrix}$$

14.21 then:

- If $(-1)^k D_k(\mathbf{x}^*) > 0$ for $k = 1, \ldots, n$, then $\mathbf{x}^*$ is a local maximum point.
- If $D_k(\mathbf{x}^*) > 0$ for $k = 1, \ldots, n$, then $\mathbf{x}^*$ is a local minimum point.
- If $D_n(\mathbf{x}^*) \neq 0$ and neither of the two conditions above is satisfied, then $\mathbf{x}^*$ is a saddle point.

Classification of stationary points of a C^2 function of n variables. *Second-order conditions* for local maximum/minimum.

$f'(x^*) = 0$ and $f''(x^*) < 0 \implies$
$\quad\quad x^*$ is a local maximum point for f.

14.22
$f'(x^*) = 0$ and $f''(x^*) > 0 \implies$
$\quad\quad x^*$ is a local minimum point for f.

One-variable second-order conditions for local maximum/minimum.

If (x_0, y_0) is a stationary point of $f(x, y)$ and $D = f_{11}''(x_0, y_0) f_{22}''(x_0, y_0) - (f_{12}''(x_0, y_0))^2$, then

- $f_{11}''(x_0, y_0) > 0$ and $D > 0 \implies$
  $\quad\quad (x_0, y_0)$ is a local minimum point for f.

14.23
- $f_{11}''(x_0, y_0) < 0$ and $D > 0 \implies$
  $\quad\quad (x_0, y_0)$ is a local maximum point for f.

- $D < 0 \implies (x_0, y_0)$ is a saddle point for f.

Two-variable second-order conditions for local maximum/minimum. (Classification of stationary points of a C^2 function of two variables.)

Optimization with equality constraints

The *Lagrange problem.* Two variables, one constraint.

14.24   max (min) $f(x,y)$ subject to $g(x,y) = b$

Lagrange's method. Recipe for solving (14.24):

(1) Introduce the *Lagrangian function*
$$\mathcal{L}(x,y) = f(x,y) - \lambda(g(x,y) - b)$$
where λ is a constant.

(2) Differentiate $\mathcal{L}$ with respect to x and y, and equate the partials to 0.

14.25

(3) The two equations in (2), together with the constraint, yield the following three equations:
$$f_1'(x,y) = \lambda g_1'(x,y)$$
$$f_2'(x,y) = \lambda g_2'(x,y)$$
$$g(x,y) = b$$

(4) Solve these three equations for the three unknowns x, y, and λ. In this way you find all possible pairs (x,y) that can solve the problem.

Necessary conditions for the solution of (14.24). Assume that $g_1'(x,y)$ and $g_2'(x,y)$ do not both vanish. For a more precise version, see (14.27). λ is called a *Lagrange multiplier*.

Suppose (x_0, y_0) satisfies the conditions in (14.25). Then:

14.26

(1) If $\mathcal{L}(x,y)$ is concave, then (x_0, y_0) solves the maximization problem in (14.24).

(2) If $\mathcal{L}(x,y)$ is convex, then (x_0, y_0) solves the minimization problem in (14.24).

Sufficient conditions for the solution of problem (14.24).

Suppose that $f(x,y)$ and $g(x,y)$ are C^1 in a domain S of the xy-plane, and that (x_0, y_0) is both an interior point of S and a local extreme point for $f(x,y)$ subject to the constraint $g(x,y) = b$.

14.27   Suppose further that $g_1'(x_0, y_0)$ and $g_2'(x_0, y_0)$ are not both 0. Then there exists a unique number λ such that the Lagrangian function
$$\mathcal{L}(x,y) = f(x,y) - \lambda(g(x,y) - b)$$
has a stationary point at (x_0, y_0).

A precise version of the Lagrange multiplier method. (*Lagrange's theorem.*)

Consider the problem

local max(min) $f(x, y)$ s.t. $g(x, y) = b$

where (x_0, y_0) satisfies the first-order conditions in (14.25). Define the bordered Hessian determinant $D(x, y)$ as

14.28

$$D(x,y) = \begin{vmatrix} 0 & g_1' & g_2' \\ g_1' & f_{11}'' - \lambda g_{11}'' & f_{12}'' - \lambda g_{12}'' \\ g_2' & f_{21}'' - \lambda g_{21}'' & f_{22}'' - \lambda g_{22}'' \end{vmatrix}$$

Local sufficient conditions for the Lagrange problem.

(1) If $D(x_0, y_0) > 0$, then (x_0, y_0) solves the local maximization problem.

(2) If $D(x_0, y_0) < 0$, then (x_0, y_0) solves the local minimization problem.

14.29

$$\max(\min) f(x_1, \ldots, x_n) \text{ s.t.} \begin{cases} g_1(x_1, \ldots, x_n) = b_1 \\ \cdots\cdots\cdots\cdots\cdots \\ g_m(x_1, \ldots, x_n) = b_m \end{cases}$$

The general *Lagrange problem.* Assume $m < n$.

Lagrange's method. Recipe for solving (14.29):

(1) Introduce the *Lagrangian function*

$$\mathcal{L}(\mathbf{x}) = f(\mathbf{x}) - \sum_{j=1}^{m} \lambda_j (g_j(\mathbf{x}) - b_j)$$

where $\lambda_1, \ldots, \lambda_m$ are constants.

14.30

(2) Equate the first-order partials of $\mathcal{L}$ to 0:

$$\frac{\partial \mathcal{L}(\mathbf{x})}{\partial x_k} = \frac{\partial f(\mathbf{x})}{\partial x_k} - \sum_{j=1}^{m} \lambda_j \frac{\partial g_j(\mathbf{x})}{\partial x_k} = 0$$

(3) Solve these n equations together with the m constraints for $x_1, \ldots, x_n$ and $\lambda_1, \ldots, \lambda_m$.

Necessary conditions for the solution of (14.29), with f and $g_1, \ldots, g_m$ as C^1 functions in an open set S in $\mathbb{R}^n$, and with $\mathbf{x} = (x_1, \ldots, x_n)$. Assume the rank of the Jacobian $(\partial g_j / \partial x_i)_{m \times n}$ to be equal to m. (See (6.8).) $\lambda_1, \ldots, \lambda_m$ are called *Lagrange multipliers.*

14.31

If $\mathbf{x}^*$ is a solution to problem (14.29) and the gradients $\nabla g_1(\mathbf{x}^*), \ldots, \nabla g_m(\mathbf{x}^*)$ are linearly independent, then there exist unique numbers $\lambda_1, \ldots, \lambda_m$ such that

$$\nabla f(\mathbf{x}^*) = \lambda_1 \nabla g_1(\mathbf{x}^*) + \cdots + \lambda_m \nabla g_m(\mathbf{x}^*)$$

An alternative formulation of (14.30).

14.32

Suppose $f(\mathbf{x})$ and $g_1(\mathbf{x}), \ldots, g_m(\mathbf{x})$ in (14.29) are defined on an open, convex set S in $\mathbb{R}^n$. Let $\mathbf{x}^* \in S$ be a stationary point of the Lagrangian and suppose $g_j(\mathbf{x}^*) = b_j$, $j = 1, \ldots, m$. Then:

$\mathcal{L}(\mathbf{x})$ concave $\Rightarrow \mathbf{x}^*$ solves problem (14.29).

Sufficient conditions for the solution of problem (14.29). (For the minimization problem, replace "$\mathcal{L}(\mathbf{x})$ concave" by "$\mathcal{L}(\mathbf{x})$ convex".)

$$14.33 \quad B_r = \begin{vmatrix} 0 & \cdots & 0 & \dfrac{\partial g_1}{\partial x_1} & \cdots & \dfrac{\partial g_1}{\partial x_r} \\ \vdots & \ddots & \vdots & \vdots & & \vdots \\ 0 & \cdots & 0 & \dfrac{\partial g_m}{\partial x_1} & \cdots & \dfrac{\partial g_m}{\partial x_r} \\ \dfrac{\partial g_1}{\partial x_1} & \cdots & \dfrac{\partial g_m}{\partial x_1} & \mathcal{L}''_{11} & \cdots & \mathcal{L}''_{1r} \\ \vdots & & \vdots & \vdots & \ddots & \vdots \\ \dfrac{\partial g_1}{\partial x_r} & \cdots & \dfrac{\partial g_m}{\partial x_r} & \mathcal{L}''_{r1} & \cdots & \mathcal{L}''_{rr} \end{vmatrix}$$

A *bordered Hessian* determinant associated with problem (14.29), $r = 1, \ldots, n$. $\mathcal{L}$ is the Lagrangian defined in (14.30).

Let $f(\mathbf{x})$ and $g_1(\mathbf{x}), \ldots, g_m(\mathbf{x})$ be C^2 functions in an open set S in $\mathbb{R}^n$, and let $\mathbf{x}^* \in S$ satisfy the necessary conditions for problem (14.29) given in (14.30). Let $B_r(\mathbf{x}^*)$ be the determinant in (14.33) evaluated at $\mathbf{x}^*$. Then:

14.34   • If $(-1)^m B_r(\mathbf{x}^*) > 0$ for $r = m+1, \ldots, n$, then $\mathbf{x}^*$ is a local minimum point for problem (14.29).

  • If $(-1)^r B_r(\mathbf{x}^*) > 0$ for $r = m+1, \ldots, n$, then $\mathbf{x}^*$ is a local maximum point for problem (14.29).

Local sufficient conditions for the Lagrange problem.

Value functions and sensitivity

14.35 $f^*(\mathbf{b}) = \max\limits_{\mathbf{x}} \{f(\mathbf{x}) : g_j(\mathbf{x}) = b_j, \ j = 1, \ldots, m\}$

$f^*(\mathbf{b})$ is the *value function*. $\mathbf{b} = (b_1, \ldots, b_m)$.

14.36 $\dfrac{\partial f^*(\mathbf{b})}{\partial b_i} = \lambda_i(\mathbf{b}), \qquad i = 1, \ldots, m$

The $\lambda_i(\mathbf{b})$'s are the unique Lagrange multipliers from (14.31). (For a precise result see Sydsæter et al. (2005), Chap. 3.)

14.37 $f^*(\mathbf{r}) = \max\limits_{\mathbf{x} \in X} f(\mathbf{x}, \mathbf{r}), \quad X \subset \mathbb{R}^n, \mathbf{r} \in A \subset \mathbb{R}^k.$

The *value function* of a maximization problem.

14.38   If $f(\mathbf{x}, \mathbf{r})$ is continuous on $X \times A$ and X is compact and nonempty, then $f^*(\mathbf{r})$ defined in (14.37) is continuous on A. If the problem in (14.37) has a unique solution $\mathbf{x} = \mathbf{x}(\mathbf{r})$ for each $\mathbf{r}$ in A, then $\mathbf{x}(\mathbf{r})$ is a continuous function of $\mathbf{r}$.

Continuity of the value function and the maximizer.

14.39   Suppose that the problem of maximizing $f(\mathbf{x}, \mathbf{r})$ for $\mathbf{x}$ in a compact set $X \subset \mathbb{R}^n$ has a unique solution $\mathbf{x}(\mathbf{r}^0)$ at $\mathbf{r} = \mathbf{r}^0$, and that $\partial f / \partial r_i$, $i = 1, \ldots, k$, exist and are continuous in a neighborhood of $(\mathbf{x}(\mathbf{r}^0), \mathbf{r}^0)$. Then for $i = 1, \ldots, k$,

$$\frac{\partial f^*(\mathbf{r}^0)}{\partial r_i} = \left[\frac{\partial f(\mathbf{x}, \mathbf{r})}{\partial r_i} \right]_{\substack{\mathbf{x} = \mathbf{x}(\mathbf{r}^0) \\ \mathbf{r} = \mathbf{r}^0}}$$

An *envelope theorem*.

14.40 $\max_{\mathbf{x}} f(\mathbf{x}, \mathbf{r})$ s.t. $g_j(\mathbf{x}, \mathbf{r}) = 0$, $j = 1, \ldots, m$

A Lagrange problem with parameters, $\mathbf{r} = (r_1, \ldots, r_k)$.

14.41 $f^*(\mathbf{r}) = \max\{f(\mathbf{x}, \mathbf{r}) : g_j(\mathbf{x}, \mathbf{r}) = 0, \; j = 1, \ldots, m\}$

The *value function* of problem (14.40).

14.42 $\dfrac{\partial f^*(\mathbf{r}^0)}{\partial r_i} = \left[\dfrac{\partial \mathcal{L}(\mathbf{x}, \mathbf{r})}{\partial r_i} \right]_{\substack{\mathbf{x} = \mathbf{x}(\mathbf{r}^0) \\ \mathbf{r} = \mathbf{r}^0}}, \quad i = 1, \ldots, k$

An *envelope theorem* for (14.40). $\mathcal{L} = f - \sum \lambda_j g_j$ is the Lagrangian. For precise assumptions for the equality to hold, see Sydsæter et al. (2005), Chapter 3.

References

See Simon and Blume (1994), Sydsæter et al. (2005), Intriligator (1971), Luenberger (1984), and Dixit (1990).

Chapter 15

Linear and nonlinear programming

Linear programming

15.1

$\max z = c_1 x_1 + \cdots + c_n x_n$ subject to

$$a_{11} x_1 + \cdots + a_{1n} x_n \leq b_1$$
$$a_{21} x_1 + \cdots + a_{2n} x_n \leq b_2$$
$$\cdots\cdots\cdots\cdots\cdots\cdots\cdots\cdots\cdots$$
$$a_{m1} x_1 + \cdots + a_{mn} x_n \leq b_m$$
$$x_1 \geq 0, \ldots, x_n \geq 0$$

A *linear programming problem*. (The *primal problem*.) $\sum_{j=1}^{n} c_j x_j$ is called the *objective function*. $(x_1, \ldots, x_n)$ is *admissible* if it satisfies all the $m + n$ constraints.

15.2

$\min Z = b_1 \lambda_1 + \cdots + b_m \lambda_m$ subject to

$$a_{11} \lambda_1 + \cdots + a_{m1} \lambda_m \geq c_1$$
$$a_{12} \lambda_1 + \cdots + a_{m2} \lambda_m \geq c_2$$
$$\cdots\cdots\cdots\cdots\cdots\cdots\cdots\cdots\cdots$$
$$a_{1n} \lambda_1 + \cdots + a_{mn} \lambda_m \geq c_n$$
$$\lambda_1 \geq 0, \ldots, \lambda_m \geq 0$$

The *dual* of (15.1). $\sum_{i=1}^{m} b_i \lambda_i$ is called the *objective function*. $(\lambda_1, \ldots, \lambda_m)$ is *admissible* if it satisfies all the $n + m$ constraints.

15.3

$\max \mathbf{c}'\mathbf{x}$ subject to $\mathbf{A}\mathbf{x} \leq \mathbf{b},\ \mathbf{x} \geq \mathbf{0}$

$\min \mathbf{b}'\boldsymbol{\lambda}$ subject to $\mathbf{A}'\boldsymbol{\lambda} \geq \mathbf{c},\ \boldsymbol{\lambda} \geq \mathbf{0}$

Matrix formulations of (15.1) and (15.2). $\mathbf{A} = (a_{ij})_{m \times n}$, $\mathbf{x} = (x_j)_{n \times 1}$, $\boldsymbol{\lambda} = (\lambda_i)_{m \times 1}$, $\mathbf{c} = (c_j)_{n \times 1}$, $\mathbf{b} = (b_i)_{m \times 1}$.

15.4

If $(x_1, \ldots, x_n)$ and $(\lambda_1, \ldots, \lambda_m)$ are admissible in (15.1) and (15.2), respectively, then

$$b_1 \lambda_1 + \cdots + b_m \lambda_m \geq c_1 x_1 + \cdots + c_n x_n$$

The value of the objective function in the dual is always greater than or equal to the value of the objective function in the primal.

15.5 Suppose $(x_1^*, \ldots, x_n^*)$ and $(\lambda_1^*, \ldots, \lambda_m^*)$ are admissible in (15.1) and (15.2) respectively, and that
$$c_1 x_1^* + \cdots + c_n x_n^* = b_1 \lambda_1^* + \cdots + b_m \lambda_m^*$$
Then $(x_1^*, \ldots, x_n^*)$ and $(\lambda_1^*, \ldots, \lambda_m^*)$ are optimal in the respective problems.

An interesting result.

15.6 If either of the problems (15.1) and (15.2) has a finite optimal solution, so has the other, and the corresponding values of the objective functions are equal. If either problem has an "unbounded optimum", then the other problem has no admissible solutions.

The *duality theorem* of linear programming.

15.7 Consider problem (15.1). If we change b_i to $b_i + \Delta b_i$ for $i = 1, \ldots, m$, and if the associated dual problem still has the same optimal solution, $(\lambda_1^*, \ldots, \lambda_m^*)$, then the change in the optimal value of the objective function of the primal problem is
$$\Delta z^* = \lambda_1^* \Delta b_1 + \cdots + \lambda_m^* \Delta b_m$$

An important sensitivity result. (The dual problem usually *will* have the same solution if $|\Delta b_1|, \ldots, |\Delta b_m|$ are sufficiently small.)

15.8 The ith optimal dual variable λ_i^* is equal to the change in objective function of the primal problem (15.1) when b_i is increased by one unit.

Interpretation of λ_i^* as a *"shadow price"*. (A special case of (15.7), with the same qualifications.)

15.9 Suppose that the primal problem (15.1) has an optimal solution $(x_1^*, \ldots, x_n^*)$ and that the dual (15.2) has an optimal solution $(\lambda_1^*, \ldots, \lambda_m^*)$. Then for $i = 1, \ldots, n$, $j = 1, \ldots, m$:

(1) $x_j^* > 0 \Rightarrow a_{1j}\lambda_1^* + \cdots + a_{mj}\lambda_m^* = c_j$

(2) $\lambda_i^* > 0 \Rightarrow a_{i1}x_1^* + \cdots + a_{in}x_n^* = b_i$

Complementary slackness. ((1): If the optimal variable j in the primal is positive, then restriction j in the dual is an equality at the optimum. (2) has a similar interpretation.)

15.10 Let $\mathbf{A}$ be an $m \times n$-matrix and $\mathbf{b}$ an n-vector. Then there exists a vector $\mathbf{y}$ with $\mathbf{Ay} \geq \mathbf{0}$ and $\mathbf{b}'\mathbf{y} < 0$ if and only if there is no $\mathbf{x} \geq \mathbf{0}$ such that $\mathbf{A}'\mathbf{x} = \mathbf{b}$.

Farkas's lemma.

Nonlinear programming

15.11 $\max \ f(x, y)$ subject to $g(x, y) \leq b$

A *nonlinear programming problem.*

Recipe for solving problem (15.11):

(1) Define the Lagrangian function $\mathcal{L}$ by
$$\mathcal{L}(x, y, \lambda) = f(x, y) - \lambda(g(x, y) - b)$$
where λ is a *Lagrange multiplier* associated with the constraint $g(x, y) \leq b$.

(2) Equate the partial derivatives of $\mathcal{L}(x, y, \lambda)$ w.r.t. x and y to zero:
$$\mathcal{L}_1'(x, y, \lambda) = f_1'(x, y) - \lambda g_1'(x, y) = 0$$
$$\mathcal{L}_2'(x, y, \lambda) = f_2'(x, y) - \lambda g_2'(x, y) = 0$$

(3) Introduce the *complementary slackness condition*
$$\lambda \geq 0 \ (\lambda = 0 \ \text{if} \ g(x, y) < b)$$

(4) Require (x, y) to satisfy $g(x, y) \leq b$.

15.12

Kuhn–Tucker necessary conditions for solving problem (15.11), made more precise in (15.20). If we find all the pairs (x, y) (together with suitable values of λ) that satisfy all these conditions, then we have all the candidates for the solution of problem. If the Lagrangian is concave in (x, y), then the conditions are sufficient for optimality.

15.13
$$\max_{\mathbf{x}} f(\mathbf{x}) \ \text{subject to} \ \begin{cases} g_1(\mathbf{x}) \leq b_1 \\ \cdots\cdots\cdots \\ g_m(\mathbf{x}) \leq b_m \end{cases}$$

A *nonlinear programming problem*. A vector $\mathbf{x} = (x_1, \ldots, x_n)$ is *admissible* if it satisfies all the constraints.

15.14
$$\mathcal{L}(\mathbf{x}, \boldsymbol{\lambda}) = f(\mathbf{x}) - \sum_{j=1}^{m} \lambda_j(g_j(\mathbf{x}) - b_j)$$

The *Lagrangian function* associated with (15.13). $\boldsymbol{\lambda} = (\lambda_1, \ldots, \lambda_m)$ are *Lagrange multipliers*.

Consider problem (15.13) and assume that f and $g_1, \ldots, g_m$ are C^1. Suppose that there exist a vector $\boldsymbol{\lambda} = (\lambda_1, \ldots, \lambda_m)$ and an admissible vector $\mathbf{x}^0 = (x_1^0, \ldots, x_n^0)$ such that

(a) $\dfrac{\partial \mathcal{L}(\mathbf{x}^0, \boldsymbol{\lambda})}{\partial x_i} = 0, \quad i = 1, \ldots, n$

15.15

(b) For all $j = 1, \ldots, m$,
$$\lambda_j \geq 0 \ (\lambda_j = 0 \ \text{if} \ g_j(\mathbf{x}^0) < b_j)$$

(c) The Lagrangian function $\mathcal{L}(\mathbf{x}, \boldsymbol{\lambda})$ is a concave function of $\mathbf{x}$.

Then $\mathbf{x}^0$ solves problem (15.13).

Sufficient conditions.

15.16 (b') $\lambda_j \geq 0$ and $\lambda_j(g_j(\mathbf{x}^0) - b_j) = 0, \ j = 1, \ldots, m$

Alternative formulation of (b) in (15.15).

15.17

(15.15) is also valid if we replace (c) by the condition

(c') $f(\mathbf{x})$ is concave and $\lambda_j g_j(\mathbf{x})$ is quasi-convex for $j = 1, \ldots, m$.

Alternative sufficient conditions.

15.18 Constraint j in (15.13) is called *active at* $\mathbf{x}^0$ if $g_j(\mathbf{x}^0) = b_j$.

Definition of an active *(or* binding*) constraint.*

15.19 The following condition is often imposed in problem (15.13): The gradients at $\mathbf{x}^0$ of those g_j-functions whose constraints are active at $\mathbf{x}^0$, are linearly independent.

A constraint qualification for problem (15.13).

15.20 Suppose that $\mathbf{x}^0 = (x_1^0, \ldots, x_n^0)$ solves (15.13) and that f and $g_1, \ldots, g_m$ are C^1. Suppose further that the constraint qualification (15.19) is satisfied at $\mathbf{x}^0$. Then there exist unique numbers $\lambda_1, \ldots, \lambda_m$ such that

(a) $\dfrac{\partial \mathcal{L}(\mathbf{x}^0, \boldsymbol{\lambda})}{\partial x_i} = 0, \quad i = 1, \ldots, n$

(b) $\lambda_j \geq 0 \; (\lambda_j = 0 \text{ if } g_j(\mathbf{x}^0) < b_j), \; j = 1, \ldots, m$

Kuhn–Tucker necessary conditions for problem (15.13). (Note that all admissible points where the constraint qualification fails to hold are candidates for optimality.)

15.21 $(\mathbf{x}^0, \boldsymbol{\lambda}^0)$ is a *saddle point* of the Lagrangian $\mathcal{L}(\mathbf{x}, \boldsymbol{\lambda})$ if
$$\mathcal{L}(\mathbf{x}, \boldsymbol{\lambda}^0) \leq \mathcal{L}(\mathbf{x}^0, \boldsymbol{\lambda}^0) \leq \mathcal{L}(\mathbf{x}^0, \boldsymbol{\lambda})$$
for all $\boldsymbol{\lambda} \geq \mathbf{0}$ and all $\mathbf{x}$.

Definition of a saddle point for problem (15.13).

15.22 If $\mathcal{L}(\mathbf{x}, \boldsymbol{\lambda})$ has a saddle point $(\mathbf{x}^0, \boldsymbol{\lambda}^0)$, then $(\mathbf{x}^0, \boldsymbol{\lambda}^0)$ solves problem (15.13).

Sufficient conditions for problem (15.13). (No differentiability or concavity conditions are required.)

15.23 The following condition is often imposed in problem (15.13): For some vector $\mathbf{x}' = (x_1', \ldots, x_n')$, $g_j(\mathbf{x}') < b_j$ for $j = 1, \ldots, m$.

The Slater condition *(constraint qualification).*

15.24 Consider problem (15.13), assuming f is concave and $g_1, \ldots, g_m$ are convex. Assume that the Slater condition (15.23) is satisfied. Then a necessary and sufficient condition for $\mathbf{x}^0$ to solve the problem is that there exist nonnegative numbers $\lambda_1^0, \ldots, \lambda_m^0$ such that $(\mathbf{x}^0, \boldsymbol{\lambda}^0)$ is a saddle point of the Lagrangian $\mathcal{L}(\mathbf{x}, \boldsymbol{\lambda})$.

A saddle point result for concave programming.

Consider problem (15.13) and assume that f and $g_1, \ldots, g_m$ are C^1. Suppose that there exist numbers $\lambda_1, \ldots, \lambda_m$ and a vector $\mathbf{x}^0$ such that

15.25
- $\mathbf{x}^0$ satisfies (a) and (b) in (15.15).
- $\nabla f(\mathbf{x}^0) \neq \mathbf{0}$
- $f(\mathbf{x})$ is quasi-concave and $\lambda_j g_j(\mathbf{x})$ is quasi-convex for $j = 1, \ldots, m$.

Then $\mathbf{x}^0$ solves problem (15.13).

Sufficient conditions for *quasi-concave programming*.

15.26 $\quad f^*(\mathbf{b}) = \max\{f(\mathbf{x}) : g_j(\mathbf{x}) \le b_j, \ j = 1, \ldots, m\}$

The *value function* of (15.13), assuming that the maximum value exists, with $\mathbf{b} = (b_1, \ldots, b_m)$.

15.27
(1) $f^*(\mathbf{b})$ is increasing in each variable.
(2) $\partial f^*(\mathbf{b})/\partial b_j = \lambda_j(\mathbf{b}), \quad j = 1, \ldots, m$
(3) If $f(\mathbf{x})$ is concave and $g_1(\mathbf{x}), \ldots, g_m(\mathbf{x})$ are convex, then $f^*(\mathbf{b})$ is concave.

Properties of the value function.

15.28 $\quad \max\limits_{\mathbf{x}} f(\mathbf{x}, \mathbf{r}) \ \text{s.t.} \ g_j(\mathbf{x}, \mathbf{r}) \le 0, \ j = 1, \ldots, m$

A nonlinear programming problem with parameters, $\mathbf{r} \in \mathbb{R}^k$.

15.29 $\quad f^*(\mathbf{r}) = \max\{f(\mathbf{x}, \mathbf{r}) : g_j(\mathbf{x}, \mathbf{r}) \le 0, \ j = 1, \ldots, m\}$

The *value function* of problem (15.28).

15.30 $\quad \dfrac{\partial f^*(\mathbf{r}^*)}{\partial r_i} = \left[\dfrac{\partial \mathcal{L}(\mathbf{x}, \mathbf{r}, \boldsymbol{\lambda})}{\partial r_i}\right]_{\substack{\mathbf{x}=\mathbf{x}(\mathbf{r}^*) \\ \mathbf{r}=\mathbf{r}^*}}, \quad i = 1, \ldots, k$

An *envelope theorem* for problem (15.28). $\mathcal{L} = f - \sum \lambda_j g_j$ is the Lagrangian. See Sydsæter et al. (2005), Section 3.8 and Clarke (1983) for a precise result.

Nonlinear programming with nonnegativity conditions

15.31 $\quad \max\limits_{\mathbf{x}} f(\mathbf{x}) \ \text{subject to} \ \begin{cases} g_1(\mathbf{x}) \le b_1 \\ \cdots\cdots\cdots \\ g_m(\mathbf{x}) \le b_m \end{cases}, \ \mathbf{x} \ge \mathbf{0}$

If the nonnegativity constraints are written as $g_{m+i}(\mathbf{x}) = -x_i \le 0$ for $i = 1, \ldots, n$, (15.31) reduces to (15.13).

Suppose in problem (15.31) that f and $g_1, \ldots,$ g_m are C^1 functions, and that there exist numbers $\lambda_1, \ldots, \lambda_m$, and an admissible vector $\mathbf{x}^0$ such that:

(a) For all $i = 1, \ldots, n$, one has $x_i^0 \geq 0$ and

$$15.32 \qquad \frac{\partial \mathcal{L}(\mathbf{x}^0, \boldsymbol{\lambda})}{\partial x_i} \leq 0, \qquad x_i^0 \frac{\partial \mathcal{L}(\mathbf{x}^0, \boldsymbol{\lambda})}{\partial x_i} = 0$$

(b) For all $j = 1, \ldots, m$,

$$\lambda_j \geq 0 \quad (\lambda_j = 0 \text{ if } g_j(\mathbf{x}^0) < b_j)$$

(c) The Lagrangian function $\mathcal{L}(\mathbf{x}, \boldsymbol{\lambda})$ is a concave function of $\mathbf{x}$.

Then $\mathbf{x}^0$ solves problem (15.31).

Sufficient conditions for problem (15.31). $\boldsymbol{\lambda} = (\lambda_i)_{m \times 1}$. $\mathcal{L}(\mathbf{x}^0, \boldsymbol{\lambda})$ is defined in (15.14).

In (15.32), (c) can be replaced by

15.33 (c') $f(\mathbf{x})$ is concave and $\lambda_j g_j(\mathbf{x})$ is quasi-convex for $j = 1, \ldots, m$.

Alternative sufficient conditions.

Suppose that $\mathbf{x}^0 = (x_1^0, \ldots, x_n^0)$ solves (15.31) and that f and $g_1, \ldots, g_m$ are C^1. Suppose also that the gradients at $\mathbf{x}^0$ of those g_j (including the functions $g_{m+1}, \ldots, g_{m+n}$ defined in the comment to (15.31)) that correspond to constraints that are active at $\mathbf{x}^0$, are linearly independent. Then there exist unique numbers $\lambda_1, \ldots, \lambda_m$ such that:

(a) For all $i = 1, \ldots, n$, $x_i^0 \geq 0$, and

$$\frac{\partial \mathcal{L}(\mathbf{x}^0, \boldsymbol{\lambda})}{\partial x_i} \leq 0, \qquad x_i^0 \frac{\partial \mathcal{L}(\mathbf{x}^0, \boldsymbol{\lambda})}{\partial x_i} = 0$$

(b) $\lambda_j \geq 0$ $(\lambda_j = 0$ if $g_j(\mathbf{x}^0) < b_j), j = 1, \ldots, m$

The Kuhn–Tucker necessary conditions for problem (15.31). (Note that all admissible points where the constraint qualification fails to hold, are candidates for optimality.)

References

Gass (1994), Luenberger (1984), Intriligator (1971), Sydsæter and Hammond (2005), Sydsæter et al. (2005), Simon and Blume (1994), Beavis and Dobbs (1990), Dixit (1990), and Clarke (1983).

Chapter 16

Calculus of variations and optimal control theory

Calculus of variations

16.1

The simplest problem in the calculus of variations (t_0, t_1, x^0, and x^1 are fixed numbers):

$$\max \int_{t_0}^{t_1} F(t, x, \dot{x})\, dt, \quad x(t_0) = x^0, \quad x(t_1) = x^1$$

F is a C^2 function. The unknown $x = x(t)$ is *admissible* if it is C^1 and satisfies the two boundary conditions. To handle the minimization problem, replace F by $-F$.

16.2

$$\frac{\partial F}{\partial x} - \frac{d}{dt}\left(\frac{\partial F}{\partial \dot{x}}\right) = 0$$

The *Euler equation*. A necessary condition for the solution of (16.1).

16.3

$$\frac{\partial^2 F}{\partial \dot{x} \partial \dot{x}} \cdot \ddot{x} + \frac{\partial^2 F}{\partial x \partial \dot{x}} \cdot \dot{x} + \frac{\partial^2 F}{\partial t \partial \dot{x}} - \frac{\partial F}{\partial x} = 0$$

An alternative form of the Euler equation.

16.4 $F''_{\dot{x}\dot{x}}(t, x(t), \dot{x}(t)) \leq 0$ for all t in $[t_0, t_1]$

The *Legendre condition*. A necessary condition for the solution of (16.1).

16.5

If $F(t, x, \dot{x})$ is concave in $(x, \dot{x})$, an admissible function $x = x(t)$ that satisfies the Euler equation, solves problem (16.1).

Sufficient conditions for the solution of (16.1).

16.6 $x(t_1)$ free in (16.1) $\Rightarrow$ $\left[\dfrac{\partial F}{\partial \dot{x}}\right]_{t=t_1} = 0$

Transversality condition. Adding condition (16.5) gives sufficient conditions.

16.7 $x(t_1) \geq x^1$ in (16.1) $\Rightarrow$

$$\left[\frac{\partial F}{\partial \dot{x}}\right]_{t=t_1} \leq 0 \quad (= 0 \text{ if } x(t_1) > x^1)$$

Transversality condition. Adding condition (16.5) gives sufficient conditions.

16.8 t_1 free in (16.1) $\Rightarrow$ $\left[F - \dot{x}\dfrac{\partial F}{\partial \dot{x}}\right]_{t=t_1} = 0$

Transversality condition.

16.9 $x(t_1) = g(t_1)$ in (16.1) $\Rightarrow$

$$\left[F + (\dot{g} - \dot{x})\frac{\partial F}{\partial \dot{x}}\right]_{t=t_1} = 0$$

Transversality condition. g is a given C^1 function.

16.10 $\max\left[\displaystyle\int_{t_0}^{t_1} F(t, x, \dot{x})\, dt + S(x(t_1))\right], \quad x(t_0) = x^0$

A variational problem with a C^1 *scrap value function, S.*

16.11 $\left[\dfrac{\partial F}{\partial \dot{x}}\right]_{t=t_1} + S'(x(t_1)) = 0$

A solution to (16.10) must satisfy (16.2) and this transversality condition.

16.12   If $F(t, x, \dot{x})$ is concave in $(x, \dot{x})$ and $S(x)$ is concave, then an admissible function satisfying the Euler equation and (16.11) solves problem (16.10).

Sufficient conditions for the solution to (16.10).

16.13 $\max \displaystyle\int_{t_0}^{t_1} F\left(t, x, \frac{dx}{dt}, \frac{d^2x}{dt^2}, \ldots, \frac{d^nx}{dt^n}\right) dt$

A variational problem with higher order derivatives. (Boundary conditions are unspecified.)

16.14 $\dfrac{\partial F}{\partial x} - \dfrac{d}{dt}\left(\dfrac{\partial F}{\partial \dot{x}}\right) + \cdots + (-1)^n \dfrac{d^n}{dt^n}\left(\dfrac{\partial F}{\partial x^{(n)}}\right) = 0$

The *(generalized) Euler equation* for (16.13).

16.15 $\max \displaystyle\iint_R F\left(t, s, x, \frac{\partial x}{\partial t}, \frac{\partial x}{\partial s}\right) dt\, ds$

A variational problem in which the unknown $x(t, s)$ is a function of two variables. (Boundary conditions are unspecified.)

16.16 $\dfrac{\partial F}{\partial x} - \dfrac{\partial}{\partial t}\left(\dfrac{\partial F}{\partial x_t'}\right) - \dfrac{\partial}{\partial s}\left(\dfrac{\partial F}{\partial x_s'}\right) = 0$

The (generalized) Euler equation for (16.15).

Optimal control theory. One state and one control variable

The simplest case. Fixed time interval $[t_0, t_1]$ and free right hand side:

16.17

$$\max \int_{t_0}^{t_1} f(t, x(t), u(t))\, dt \quad u(t) \in \mathbb{R},$$

$$\dot{x}(t) = g(t, x(t), u(t)), \quad x(t_0) = x^0, \quad x(t_1) \text{ free}$$

The pair $(x(t), u(t))$ is *admissible* if it satisfies the differential equation, $x(t_0) = x^0$, and $u(t)$ is piecewise continuous. To handle the minimization problem, replace f by $-f$.

16.18 $\quad H(t, x, u, p) = f(t, x, u) + pg(t, x, u)$

The Hamiltonian associated with (16.17).

16.19

Suppose $(x^*(t), u^*(t))$ solves problem (16.17). Then there exists a continuous function $p(t)$ such that for each t in $[t_0, t_1]$,

(1) $H(t, x^*(t), u, p(t)) \leq H(t, x^*(t), u^*(t), p(t))$
     for all u in $\mathbb{R}$. In particular,

     $H'_u(t, x^*(t), u^*(t), p(t)) = 0$

(2) The function $p(t)$ satisfies

     $\dot{p}(t) = -H'_x(t, x^*(t), u^*(t), p(t)), \quad p(t_1) = 0$

The *maximum principle.* The differential equation for $p(t)$ is not necessarily valid at the discontinuity points of $u^*(t)$. The equation $p(t_1) = 0$ is called a *transversality condition.*

16.20

If $(x^*(t), u^*(t))$ satisfies the conditions in (16.19) and $H(t, x, u, p(t))$ is concave in (x, u), then $(x^*(t), u^*(t))$ solves problem (16.17).

Mangasarian's sufficient conditions for problem (16.17).

16.21

$$\max \int_{t_0}^{t_1} f(t, x(t), u(t))\, dt, \quad u(t) \in U \subset \mathbb{R},$$

$$\dot{x}(t) = g(t, x(t), u(t)), \quad x(t_0) = x^0$$

(a) $x(t_1) = x^1 \quad$ or $\quad$ (b) $x(t_1) \geq x^1$

A control problem with terminal conditions and fixed time interval. U is the *control region.* $u(t)$ is piecewise continuous.

16.22 $\quad H(t, x, u, p) = p_0 f(t, x, u) + pg(t, x, u)$

The Hamiltonian associated with (16.21).

16.23

Suppose $(x^*(t), u^*(t))$ solves problem (16.21). Then there exist a continuous function $p(t)$ and a number p_0 such that for all t in $[t_0, t_1]$,

(1) $p_0 = 0$ or 1 and $(p_0, p(t))$ is never $(0, 0)$.

(2) $H(t, x^*(t), u, p(t)) \leq H(t, x^*(t), u^*(t), p(t))$
     for all u in U.

(3) $\dot{p}(t) = -H'_x(t, x^*(t), u^*(t), p(t))$

(4) (a') No conditions on $p(t_1)$.

     (b') $p(t_1) \geq 0 \quad (p(t_1) = 0$ if $x^*(t_1) > x^1)$

The *maximum principle.* The differential equation for $p(t)$ is not necessarily valid at the discontinuity points of $u^*(t)$. (4)(b') is called a *transversality condition.* (Except in degenerate cases, one can put $p_0 = 1$ and then ignore (1).)

Several state and control variables

$$\max \int_{t_0}^{t_1} f(t, \mathbf{x}(t), \mathbf{u}(t))\, dt$$

$$\dot{\mathbf{x}}(t) = \mathbf{g}(t, \mathbf{x}(t), \mathbf{u}(t)), \quad \mathbf{x}(t_0) = \mathbf{x}^0$$

16.24 $\quad \mathbf{u}(t) = (u_1(t), \ldots, u_r(t)) \in U \subset \mathbb{R}^r$

(a) $x_i(t_1) = x_i^1, \qquad i = 1, \ldots, l$

(b) $x_i(t_1) \geq x_i^1, \qquad i = l+1, \ldots, q$

(c) $x_i(t_1)$ free, $\qquad i = q+1, \ldots, n$

> A standard control problem with fixed time interval. U is the control region, $\mathbf{x}(t) = (x_1(t), \ldots, x_n(t))$, $\mathbf{g} = (g_1, \ldots, g_n)$. $\mathbf{u}(t)$ is piecewise continuous.

16.25 $\quad H(t, \mathbf{x}, \mathbf{u}, \mathbf{p}) = p_0 f(t, \mathbf{x}, \mathbf{u}) + \sum_{i=1}^{n} p_i g_i(t, \mathbf{x}, \mathbf{u})$

> The *Hamiltonian*.

If $(\mathbf{x}^*(t), \mathbf{u}^*(t))$ solves problem (16.24), there exist a constant p_0 and a continuous function $\mathbf{p}(t) = (p_1(t), \ldots, p_n(t))$, such that for all t in $[t_0, t_1]$,

(1) $p_0 = 0$ or 1 and $(p_0, \mathbf{p}(t))$ is never $(0, \mathbf{0})$.

(2) $H(t, \mathbf{x}^*(t), \mathbf{u}, \mathbf{p}(t)) \leq H(t, \mathbf{x}^*(t), \mathbf{u}^*(t), \mathbf{p}(t))$ for all $\mathbf{u}$ in U.

16.26

(3) $\dot{p}_i(t) = -\partial H^*/\partial x_i, \qquad i = 1, \ldots, n$

(4) (a') No condition on $p_i(t_1), \qquad i = 1, \ldots, l$

(b') $p_i(t_1) \geq 0 \quad (= 0 \text{ if } x_i^*(t_1) > x_i^1)$
$\qquad\qquad\qquad\qquad\qquad i = l+1, \ldots, q$

(c') $p_i(t_1) = 0, \qquad\qquad i = q+1, \ldots, n$

> The *maximum principle*. H^* denotes evaluation at $(t, \mathbf{x}^*(t), \mathbf{u}^*(t), \mathbf{p}(t))$. The differential equation for $p_i(t)$ is not necessarily valid at the discontinuity points of $\mathbf{u}^*(t)$. (4) (b') and (c') are *transversality conditions*. (Except in degenerate cases, one can put $p_0 = 1$ and then ignore (1).)

If $(\mathbf{x}^*(t), \mathbf{u}^*(t))$ satisfies all the conditions in (16.26) for $p_0 = 1$, and $H(t, \mathbf{x}, \mathbf{u}, \mathbf{p}(t))$ is concave in $(\mathbf{x}, \mathbf{u})$, then $(\mathbf{x}^*(t), \mathbf{u}^*(t))$ solves problem (16.24).

16.27

> *Mangasarian's sufficient conditions* for problem (16.24).

The condition in (16.27) that $H(t, \mathbf{x}, \mathbf{u}, \mathbf{p}(t))$ is concave in $(\mathbf{x}, \mathbf{u})$, can be replaced by the weaker condition that the *maximized Hamiltonian*

16.28

$$\hat{H}(t, \mathbf{x}, \mathbf{p}(t)) = \max_{\mathbf{u} \in U} H(t, \mathbf{x}, \mathbf{u}, \mathbf{p}(t))$$

is concave in $\mathbf{x}$.

> *Arrow's sufficient condition*.

16.29 $\quad V(\mathbf{x}^0, \mathbf{x}^1, t_0, t_1) = \int_{t_0}^{t_1} f(t, \mathbf{x}^*(t), \mathbf{u}^*(t))\, dt$

> The *value function* of problem (16.24), assuming that the solution is $(\mathbf{x}^*(t), \mathbf{u}^*(t))$ and that $\mathbf{x}^1 = (x_1^1, \ldots, x_q^1)$.

16.30

$$\frac{\partial V}{\partial x_i^0} = p_i(t_0), \quad i = 1, \ldots, n$$

$$\frac{\partial V}{\partial x_i^1} = -p_i(t_1), \quad i = 1, \ldots, q$$

$$\frac{\partial V}{\partial t_0} = -H^*(t_0), \quad \frac{\partial V}{\partial t_1} = H^*(t_1)$$

Properties of the value function, assuming V is differentiable. $H^*(t) = H(t, \mathbf{x}^*(t), \mathbf{u}^*(t), \mathbf{p}(t))$. (For precise assumptions, see Seierstad and Sydsæter (1987), Sec. 3.5.)

16.31

If t_1 is free in problem (16.24) and $(\mathbf{x}^*(t), \mathbf{u}^*(t))$ solves the corresponding problem on $[t_0, t_1^*]$, then all the conditions in (16.26) are satisfied on $[t_0, t_1^*]$, and in addition

$$H(t_1^*, \mathbf{x}^*(t_1^*), \mathbf{u}^*(t_1^*), \mathbf{p}(t_1^*)) = 0$$

Necessary conditions for a free terminal time problem. (Concavity of the Hamiltonian in $(\mathbf{x}, \mathbf{u})$ is *not* sufficient for optimality when t_1 is free. See Seierstad and Sydsæter (1987), Sec. 2.9.)

16.32

Replace the terminal conditions (a), (b), and (c) in problem (16.24) by

$$R_k(\mathbf{x}(t_1)) = 0, \quad k = 1, 2, \ldots, r_1',$$

$$R_k(\mathbf{x}(t_1)) \geq 0, \quad k = r_1' + 1, r_1' + 2, \ldots, r_1,$$

where $R_1, \ldots, R_{r_1}$ are C^1 functions. If the pair $(\mathbf{x}^*(t), \mathbf{u}^*(t))$ is optimal, then the conditions in (16.26) are satisfied, except that (4) is replaced by the condition that there exist numbers $a_1, \ldots, a_{r_1}$ such that

$$p_j(t_1) = \sum_{k=1}^{r_1} a_k \frac{\partial R_k(\mathbf{x}^*(t_1))}{\partial x_j}, \quad j = 1, \ldots, n$$

where $a_k \geq 0$ ($a_k = 0$ if $R_k(\mathbf{x}^*(t_1)) > 0$) for $k = r_1' + 1, \ldots, r_1$, and (1) is replaced by

$$p_0 = 0 \text{ or } 1, \; (p_0, a_1, \ldots, a_{r_1}) \neq (0, 0, \ldots, 0)$$

If $\widehat{H}(t, \mathbf{x}, \mathbf{p}(t))$ is concave in $\mathbf{x}$ for $p_0 = 1$ and the sum $\sum_{k=1}^{r_1} a_k R_k(\mathbf{x})$ is quasi-concave in $\mathbf{x}$, then $(\mathbf{x}^*, \mathbf{u}^*)$ is optimal.

More general terminal conditions. $\widehat{H}(t, \mathbf{x}, \mathbf{p}(t))$ is defined in (16.28).

16.33

$$\max\left[\int_{t_0}^{t_1} f(t, \mathbf{x}(t), \mathbf{u}(t)) e^{-rt} \, dt + S(t_1, \mathbf{x}(t_1)) e^{-rt_1} \right]$$

$$\dot{\mathbf{x}}(t) = \mathbf{g}(t, \mathbf{x}(t), \mathbf{u}(t)), \quad \mathbf{x}(t_0) = \mathbf{x}^0, \quad \mathbf{u}(t) \in U \subset \mathbb{R}^r$$

(a) $x_i(t_1) = x_i^1, \quad i = 1, \ldots, l$

(b) $x_i(t_1) \geq x_i^1, \quad i = l + 1, \ldots, q$

(c) $x_i(t_1)$ free, $\quad i = q + 1, \ldots, n$

A control problem with a *scrap value function*, S. t_0 and t_1 are fixed.

16.34 $H^c(t, \mathbf{x}, \mathbf{u}, \mathbf{q}) = q_0 f(t, \mathbf{x}, \mathbf{u}) + \sum_{j=1}^{n} q_j g_j(t, \mathbf{x}, \mathbf{u})$

> The *current value Hamiltonian* for problem (16.33).

16.35

If $(\mathbf{x}^*(t), \mathbf{u}^*(t))$ solves problem (16.33), there exists a constant q_0 and a continuous function $\mathbf{q}(t) = (q_1(t), \ldots, q_n(t))$ such that for all t in $[t_0, t_1]$,

(1) $q_0 = 0$ or 1 and $(q_0, \mathbf{q}(t))$ is never $(0, \mathbf{0})$.

(2) $H^c(t, \mathbf{x}^*(t), \mathbf{u}, \mathbf{q}(t)) \leq H^c(t, \mathbf{x}^*(t), \mathbf{u}^*(t), \mathbf{q}(t))$ for all $\mathbf{u}$ in U.

(3) $\dot{q}_i - rq_i = -\dfrac{\partial H^c(t, \mathbf{x}^*, \mathbf{u}^*, \mathbf{q})}{\partial x_i}$, $\quad i = 1, \ldots, n$

(4)

(a') No condition on $q_i(t_1)$, $\qquad i = 1, \ldots, l$

(b') $q_i(t_1) \geq q_0 \dfrac{\partial S^*(t_1, \mathbf{x}^*(t_1))}{\partial x_i}$

$\qquad$ (with = if $x_i^*(t_1) > x_i^1$), $\qquad i = l+1, \ldots, m$

(c') $q_i(t_1) = q_0 \dfrac{\partial S^*(t_1, \mathbf{x}^*(t_1))}{\partial x_i}$, $\quad i = m+1, \ldots, n$

> The maximum principle for problem (16.33), *current value formulation*. The differential equation for $q_i = q_i(t)$ is not necessarily valid at the discontinuity points of $\mathbf{u}^*(t)$. (Except in degenerate cases, one can put $q_0 = 1$ and then ignore (1).)

16.36

If $(\mathbf{x}^*(t), \mathbf{u}^*(t))$ satisfies the conditions in (16.35) for $q_0 = 1$, if $H^c(t, \mathbf{x}, \mathbf{u}, \mathbf{q}(t))$ is concave in $(\mathbf{x}, \mathbf{u})$, and if $S(t, \mathbf{x})$ is concave in $\mathbf{x}$, then the pair $(\mathbf{x}^*(t), \mathbf{u}^*(t))$ solves the problem.

> Sufficient conditions for the solution of (16.33). (Mangasarian.)

16.37

The condition in (16.36) that $H^c(t, \mathbf{x}, \mathbf{u}, \mathbf{q}(t))$ is concave in $(\mathbf{x}, \mathbf{u})$ can be replaced by the weaker condition that the *maximized current value Hamiltonian*

$$\widehat{H}^c(t, \mathbf{x}, \mathbf{q}(t)) = \max_{\mathbf{u} \in U} H^c(t, \mathbf{x}, \mathbf{u}, \mathbf{q}(t))$$

is concave in $\mathbf{x}$.

> *Arrow's sufficient condition.*

16.38

If t_1 is free in problem (16.33), and if $(\mathbf{x}^*, \mathbf{u}^*)$ solves the corresponding problem on $[t_0, t_1^*]$, then all the conditions in (16.35) are satisfied on $[t_0, t_1^*]$, and in addition

$$H^c(t_1^*, \mathbf{x}^*(t_1^*), \mathbf{u}^*(t_1^*), \mathbf{q}(t_1^*)) =$$

$$q_0 r S(t_1^*, \mathbf{x}^*(t_1^*)) - q_0 \frac{\partial S(t_1^*, \mathbf{x}^*(t_1^*))}{\partial t_1}$$

> Necessary conditions for problem (16.33) when t_1 is free. (Except in degenerate cases, one can put $q_0 = 1$.)

Linear quadratic problems

16.39

$$\min\Big[\int_{t_0}^{t_1} (\mathbf{x}'\mathbf{A}\mathbf{x} + \mathbf{u}'\mathbf{B}\mathbf{u})\,dt + (\mathbf{x}(t_1))'\mathbf{S}\mathbf{x}(t_1)\Big],$$

$$\dot{\mathbf{x}} = \mathbf{F}\mathbf{x} + \mathbf{G}\mathbf{u}, \quad \mathbf{x}(t_0) = \mathbf{x}^0, \quad \mathbf{u} \in \mathbb{R}^r.$$

The matrices $\mathbf{A} = \mathbf{A}(t)_{n\times n}$ and $\mathbf{S}_{n\times n}$ are symmetric and positive semidefinite, $\mathbf{B} = \mathbf{B}(t)_{r\times r}$ is symmetric and positive definite, $\mathbf{F} = \mathbf{F}(t)_{n\times n}$ and $\mathbf{G} = \mathbf{G}(t)_{n\times r}$.

A linear quadratic control problem. The entries of $\mathbf{A}(t)$, $\mathbf{B}(t)$, $\mathbf{F}(t)$, and $\mathbf{G}(t)$ are continuous functions of t. $\mathbf{x} = \mathbf{x}(t)$ is $n \times 1$, $\mathbf{u} = \mathbf{u}(t)$ is $r \times 1$.

16.40

$$\dot{\mathbf{R}} = -\mathbf{R}\mathbf{F} - \mathbf{F}'\mathbf{R} + \mathbf{R}\mathbf{G}\mathbf{B}^{-1}\mathbf{G}'\mathbf{R} - \mathbf{A}$$

The *Riccati* equation associated with (16.39).

16.41

Suppose $(\mathbf{x}^*(t), \mathbf{u}^*(t))$ is admissible in problem (16.39), and let $\mathbf{u}^* = -(\mathbf{B}(t))^{-1}(\mathbf{G}(t))'\mathbf{R}(t)\mathbf{x}^*$, with $\mathbf{R} = \mathbf{R}(t)$ as a symmetric $n \times n$-matrix with C^1 entries satisfying (16.40) with $\mathbf{R}(t_1) = \mathbf{S}$. Then $(\mathbf{x}^*(t), \mathbf{u}^*(t))$ solves problem (16.39).

The solution of (16.39).

Infinite horizon

16.42

$$\max \int_{t_0}^{\infty} f(t, x(t), u(t))e^{-rt}\,dt,$$

$$\dot{x}(t) = g(t, x(t), u(t)), \quad x(t_0) = x^0, \quad u(t) \in U,$$

$$\lim_{t\to\infty} x(t) \geq x^1 \quad (x^1 \text{ is a fixed number}).$$

A simple one-variable *infinite horizon problem*, assuming that the integral converges for all admissible pairs.

16.43

$$H^c(t, x, u, q) = q_0 f(t, x, u) + q g(t, x, u)$$

The current value Hamiltonian for problem (16.42).

16.44

Suppose that, with $q_0 = 1$, an admissible pair $(x^*(t), u^*(t))$ in problem (16.42) satisfies the following conditions for all $t \geq t_0$:

(1) $H^c(t, x^*(t), u, q(t)) \leq H^c(t, x^*(t), u^*(t), q(t))$ for all u in U.

(2) $\dot{q}(t) - rq = -\partial H^c(t, x^*(t), u^*(t), q(t))/\partial x$

(3) $H^c(t, x, u, q(t))$ is concave in (x, u).

(4) $\lim_{t\to\infty}[q(t)e^{-rt}(x(t) - x^*(t))] \geq 0$ for all admissible $x(t)$.

Then $(x^*(t), u^*(t))$ is optimal.

Mangasarian's sufficient conditions. (Conditions (1) and (2) are (essentially) necessary for problem (16.42), but (4) is not. For a discussion of necessary conditions, see e.g. Seierstad and Sydsæter (1987), Sec. 3.7.)

16.45

$$\max \int_{t_0}^{\infty} f(t, \mathbf{x}(t), \mathbf{u}(t)) e^{-rt}\, dt$$

$$\dot{\mathbf{x}}(t) = \mathbf{g}(t, \mathbf{x}(t), \mathbf{u}(t)), \quad \mathbf{x}(t_0) = \mathbf{x}^0, \quad \mathbf{u}(t) \in U \subset \mathbb{R}^r$$

(a) $\underline{\lim}_{t \to \infty} x_i(t) = x_i^1, \qquad i = 1, \ldots, l$

(b) $\underline{\lim}_{t \to \infty} x_i(t) \geq x_i^1, \qquad i = l+1, \ldots, m$

(c) $x_i(t)$ free as $t \to \infty$, $\qquad i = m+1, \ldots, n$

An infinite horizon problem with several state and control variables. For $\underline{\lim}$, see (12.42) and (12.43).

16.46

$$D(t) = \int_{t_0}^{t} (f^* - f) e^{-r\tau}\, d\tau, \quad \text{where}$$

$$f^* = f(\tau, \mathbf{x}^*(\tau), \mathbf{u}^*(\tau)), \quad f = f(\tau, \mathbf{x}(\tau), \mathbf{u}(\tau))$$

Notation for (16.47). $(\mathbf{x}^*(t), \mathbf{u}^*(t))$ is a candidate for optimality, and $(\mathbf{x}(t), \mathbf{u}(t))$ is any admissible pair.

16.47

The pair $(\mathbf{x}^*(t), \mathbf{u}^*(t))$ is

- *sporadically catching up optimal* (SCU-optimal) if for every admissible pair $(\mathbf{x}(t), \mathbf{u}(t))$,

  $$\overline{\lim}_{t \to \infty} D(t) \geq 0$$

  i.e. for every $\varepsilon > 0$ and every T there is some $t \geq T$ such that $D(t) \geq -\varepsilon$;

- *catching up optimal* (CU-optimal) if for every admissible pair $(\mathbf{x}(t), \mathbf{u}(t))$,

  $$\underline{\lim}_{t \to \infty} D(t) \geq 0$$

  i.e. for every $\varepsilon > 0$ there exists a T such that $D(t) \geq -\varepsilon$ for all $t \geq T$;

- *overtaking optimal* (OT-optimal) if for every admissible pair $(\mathbf{x}(t), \mathbf{u}(t))$, there exists a number T such that $D(t) \geq 0$ for all $t \geq T$.

Different optimality criteria for infinite horizon problems. For $\underline{\lim}$ and $\overline{\lim}$, see (12.42) and (12.43). (SCU-optimality is also called weak optimality, while CU-optimality is then called overtaking optimality.)

16.48

OT-optimality $\Rightarrow$ CU-optimality

$\Rightarrow$ SCU-optimality

Relationship between the optimality criteria.

16.49

Suppose $(\mathbf{x}^*(t), \mathbf{u}^*(t))$ is SCU-, CU-, or OT-optimal in problem (16.45). Then there exist a constant q_0 and a continuous function $\mathbf{q}(t) = (q_1(t), \ldots, q_n(t))$ such that for all $t \geq t_0$,

(1) $q_0 = 0$ or 1 and $(q_0, \mathbf{q}(t))$ is never $(0, \mathbf{0})$.

(2) $H^c(t, \mathbf{x}^*(t), \mathbf{u}, \mathbf{q}(t)) \leq H^c(t, \mathbf{x}^*(t), \mathbf{u}^*(t), \mathbf{q}(t))$ for all $\mathbf{u}$ in U.

(3) $\dot{q}_i - r q_i = -\dfrac{\partial H^c(t, \mathbf{x}^*, \mathbf{u}^*, \mathbf{q})}{\partial x_i}, \quad i = 1, \ldots, n$

The maximum principle. Infinite horizon. (No transversality condition.) The differential equation for $q_i(t)$ is not necessarily valid at the discontinuity points of $\mathbf{u}^*(t)$.

With regard to CU-optimality, conditions (2) and (3) in (16.49) (with $q_0 = 1$) are sufficient for optimality if

16.50

(1) $H^c(t, \mathbf{x}, \mathbf{u}, \mathbf{q}(t))$ is concave in $(\mathbf{x}, \mathbf{u})$

(2) $\underline{\lim}_{t \to \infty} e^{-rt} \mathbf{q}(t) \cdot (\mathbf{x}(t) - \mathbf{x}^*(t)) \geq 0$ for all admissible $\mathbf{x}(t)$.

Sufficient conditions for the infinite horizon case.

Condition (16.50) (2) is satisfied if the following conditions are satisfied for all admissible $\mathbf{x}(t)$:

(1) $\lim_{t \to \infty} e^{-rt} q_i(t)(x_i^1 - x_i^*(t)) \geq 0$, $i = 1, \ldots, m$.

(2) There exists a constant M such that $|e^{-rt} q_i(t)| \leq M$ for all $t \geq t_0$, $i = 1, \ldots, m$.

16.51

(3) Either there exists a number $t' \geq t_0$ such that $q_i(t) \geq 0$ for all $t \geq t'$, or there exists a number P such that $|x_i(t)| \leq P$ for all $t \geq t_0$ and $\underline{\lim}_{t \to \infty} q_i(t) \geq 0$, $i = l + 1, \ldots, m$.

(4) There exists a number Q such that $|x_i(t)| < Q$ for all $t \geq t_0$, and $\lim_{t \to \infty} q_i(t) = 0$, $i = m + 1, \ldots, n$.

Sufficient conditions for (16.50) (2) to hold. See Seierstad and Sydsæter (1987), Section 3.7, Note 16.

Mixed constraints

16.52

$$\max \int_{t_0}^{t_1} f(t, \mathbf{x}(t), \mathbf{u}(t)) \, dt$$

$$\dot{\mathbf{x}}(t) = \mathbf{g}(t, \mathbf{x}(t), \mathbf{u}(t)), \quad \mathbf{x}(t_0) = \mathbf{x}^0, \quad \mathbf{u}(t) \in \mathbb{R}^r$$

$$h_k(t, \mathbf{x}(t), \mathbf{u}(t)) \geq 0, \quad k = 1, \ldots, s$$

(a) $x_i(t_1) = x_i^1$, $\quad i = 1, \ldots, l$

(b) $x_i(t_1) \geq x_i^1$, $\quad i = l + 1, \ldots, q$

(c) $x_i(t_1)$ free, $\quad i = q + 1, \ldots, n$

A mixed constraints problem. $\mathbf{x}(t) \in \mathbb{R}^n$. $h_1, \ldots, h_s$ are given functions. (All restrictions on $\mathbf{u}$ must be included in the h_k constraints.)

16.53 $\quad \mathcal{L}(t, \mathbf{x}, \mathbf{u}, \mathbf{p}, \mathbf{q}) = H(t, \mathbf{x}, \mathbf{u}, \mathbf{p}) + \sum_{k=1}^{s} q_k h_k(t, \mathbf{x}, \mathbf{u})$

The Lagrangian associated with (16.52). $H(t, \mathbf{x}, \mathbf{u}, \mathbf{p})$ is the usual Hamiltonian.

Suppose $(\mathbf{x}^*(t), \mathbf{u}^*(t))$ is an admissible pair in problem (16.52). Suppose further that there exist functions $\mathbf{p}(t) = (p_1(t), \ldots, p_n(t))$ and $\mathbf{q}(t) = (q_1(t), \ldots, q_s(t))$, where $\mathbf{p}(t)$ is continuous and $\dot{\mathbf{p}}(t)$ and $\mathbf{q}(t)$ are piecewise continuous, such that the following conditions are satisfied with $p_0 = 1$:

16.54

(1) $\dfrac{\partial \mathcal{L}^*}{\partial u_j} = 0$, $\qquad\qquad j = 1, \ldots, r$

(2) $q_k(t) \geq 0$ $(q_k(t) = 0$ if $h_k(t, \mathbf{x}^*(t), \mathbf{u}^*(t)) > 0)$, $\qquad\qquad k = 1, \ldots, s$

(3) $\dot{p}_i(t) = -\dfrac{\partial \mathcal{L}^*}{\partial x_i}$, $\qquad\qquad i = 1, \ldots, n$

(4)

(a') No conditions on $p_i(t_1)$, $\qquad i = 1, \ldots, l$

(b') $p_i(t_1) \geq 0$ $(p_i(t_1) = 0$ if $x_i^*(t_1) > x_i^1)$, $\qquad\qquad i = l+1, \ldots, m$

(c') $p_i(t_1) = 0$, $\qquad\qquad i = m+1, \ldots, n$

(5) $H(t, \mathbf{x}, \mathbf{u}, \mathbf{p}(t))$ is concave in $(\mathbf{x}, \mathbf{u})$

(6) $h_k(t, \mathbf{x}, \mathbf{u})$ is quasi-concave in $(\mathbf{x}, \mathbf{u})$, $\qquad\qquad k = 1, \ldots, s$

Then $(\mathbf{x}^*(t), \mathbf{u}^*(t))$ solves the problem.

Mangasarian's sufficient conditions for problem (16.52). $\mathcal{L}^*$ denotes evaluation at $(t, \mathbf{x}^*(t), \mathbf{u}^*(t), \mathbf{p}(t), \mathbf{q}(t))$. (The standard *necessary* conditions for optimality involve a constraint qualification that severely restricts the type of functions that can appear in the h_k-constraints. In particular, each constraint active at the optimum must contain at least one of the control variables as an argument. For details, see the references.)

Pure state constraints

16.55

$$\max \int_{t_0}^{t_1} f(t, \mathbf{x}(t), \mathbf{u}(t))\, dt$$

$$\dot{\mathbf{x}}(t) = \mathbf{g}(t, \mathbf{x}(t), \mathbf{u}(t)), \quad \mathbf{x}(t_0) = \mathbf{x}^0$$

$$\mathbf{u}(t) = (u_1(t), \ldots, u_r(t)) \in U \subset \mathbb{R}^r$$

$$h_k(t, \mathbf{x}(t)) \geq 0, \quad k = 1, \ldots, s$$

(a) $x_i(t_1) = x_i^1$, $\qquad i = 1, \ldots, l$

(b) $x_i(t_1) \geq x_i^1$, $\qquad i = l+1, \ldots, q$

(c) $x_i(t_1)$ free, $\qquad i = q+1, \ldots, n$

A pure state constraints problem. U is the control region. $h_1, \ldots, h_s$ are given functions.

16.56 $\mathcal{L}(t, \mathbf{x}, \mathbf{u}, \mathbf{p}, \mathbf{q}) = H(t, \mathbf{x}, \mathbf{u}, \mathbf{p}) + \displaystyle\sum_{k=1}^{s} q_k h_k(t, \mathbf{x})$

The Lagrangian associated with (16.55). $H(t, \mathbf{x}, \mathbf{u}, \mathbf{p})$ is the usual Hamiltonian.

Suppose $(\mathbf{x}^*(t), \mathbf{u}^*(t))$ is admissible in problem (16.55), and that there exist vector functions $\mathbf{p}(t)$ and $\mathbf{q}(t)$, where $\mathbf{p}(t)$ is continuous and $\dot{\mathbf{p}}(t)$ and $\mathbf{q}(t)$ are piecewise continuous in $[t_0, t_1)$, and numbers β_k, $k = 1, \ldots, s$, such that the following conditions are satisfied with $p_0 = 1$:

(1) $\mathbf{u} = \mathbf{u}^*(t)$ maximizes $H(t, \mathbf{x}^*(t), \mathbf{u}, \mathbf{p}(t))$ for $\mathbf{u}$ in U.

(2) $q_k(t) \geq 0$ $(q_k(t) = 0$ if $h_k(t, \mathbf{x}^*(t)) > 0)$,
$$k = 1, \ldots, s$$

(3) $\dot{p}_i(t) = -\dfrac{\partial \mathcal{L}^*}{\partial x_i}$, $\qquad i = 1, \ldots, n$

(4) At t_1, $p_i(t)$ can have a jump discontinuity, in which case
$$p_i(t_1^-) - p_i(t_1) = \sum_{k=1}^{s} \beta_k \frac{\partial h_k(t_1, \mathbf{x}^*(t_1))}{\partial x_i},$$
$$i = 1, \ldots, n$$

16.57

(5) $\beta_k \geq 0$ $(\beta_k = 0$ if $h_k(t_1, \mathbf{x}^*(t_1)) > 0)$,
$$k = 1, \ldots, s$$

(6) (a') No conditions on $p_i(t_1)$, $\qquad i = 1, \ldots, l$

(b') $p_i(t_1) \geq 0$ $(p_i(t_1) = 0$ if $x_i^*(t_1) > x_i^1)$,
$$i = l+1, \ldots, m$$

(c') $p_i(t_1) = 0$, $\qquad i = m+1, \ldots, n$

(7) $\widehat{H}(t, \mathbf{x}, \mathbf{p}(t)) = \max_{\mathbf{u} \in U} H(t, \mathbf{x}, \mathbf{u}, \mathbf{p}(t))$ is concave in $\mathbf{x}$.

(8) $h_k(t, \mathbf{x})$ is quasi-concave in $\mathbf{x}$, $\qquad k = 1, \ldots, s$

Then $(\mathbf{x}^*(t), \mathbf{u}^*(t))$ solves the problem.

Mangasarian's sufficient conditions for the pure state constraints problem (16.55). $\mathbf{p}(t) = (p_1(t), \ldots, p_n(t))$ and $\mathbf{q}(t) = (q_1(t), \ldots, q_s(t))$. $\mathcal{L}^*$ denotes evaluation at $(t, \mathbf{x}^*(t), \mathbf{u}^*(t), \mathbf{p}(t), \mathbf{q}(t))$. (The conditions in the theorem are somewhat restrictive. In particular, sometimes one must allow $\mathbf{p}(t)$ to have discontinuities at interior points of $[t_0, t_1]$. For details, see the references.)

Mixed and pure state constraints

$$\max \int_{t_0}^{t_1} f(t, \mathbf{x}(t), \mathbf{u}(t))\, dt$$

$\dot{\mathbf{x}}(t) = \mathbf{g}(t, \mathbf{x}(t), \mathbf{u}(t))$, $\quad \mathbf{x}(t_0) = \mathbf{x}^0$

$\mathbf{u}(t) = (u_1(t), \ldots, u_r(t)) \in U \subset \mathbb{R}^r$

16.58 $\quad h_k(t, \mathbf{x}(t), \mathbf{u}(t)) \geq 0$, $\qquad k = 1, \ldots, s'$

$h_k(t, \mathbf{x}(t), \mathbf{u}(t)) = \bar{h}_k(t, \mathbf{x}(t)) \geq 0$, $\ k = s'+1, \ldots, s$

(a) $x_i(t_1) = x_i^1$, $\qquad i = 1, \ldots, l$

(b) $x_i(t_1) \geq x_i^1$, $\qquad i = l+1, \ldots, q$

(c) $x_i(t_1)$ free, $\qquad i = q+1, \ldots, n$

A mixed and pure state constraints problem.

Let $(\mathbf{x}^*(t), \mathbf{u}^*(t))$ be admissible in problem (16.58). Assume that there exist vector functions $\mathbf{p}(t)$ and $\mathbf{q}(t)$, where $\mathbf{p}(t)$ is continuous and $\dot{\mathbf{p}}(t)$ and $\mathbf{q}(t)$ are piecewise continuous, and also numbers β_k, $k = 1, \ldots, s$, such that the following conditions are satisfied with $p_0 = 1$:

16.59

(1) $\left(\dfrac{\partial \mathcal{L}^*}{\partial \mathbf{u}} \right) \cdot (\mathbf{u} - \mathbf{u}^*(t)) \leq 0$ for all $\mathbf{u}$ i U

(2) $\dot{p}_i(t) = -\dfrac{\partial \mathcal{L}^*}{\partial x_i}$, $i = 1, \ldots, n$

(3) $p_i(t_1) - \sum\limits_{k=1}^{s} \beta_k \dfrac{\partial h_k(t_1, \mathbf{x}^*(t_1), \mathbf{u}^*(t_1))}{\partial x_i}$

    satisfies

    (a') no conditions, $i = 1, \ldots, l$

    (b') ≥ 0 $(= 0$ if $x_i^*(t_1) > x_i^1)$,

                    $i = l+1, \ldots, m$

    (c') $= 0$, $i = m+1, \ldots, n$

(4) $\beta_k = 0$, $k = 1, \ldots, s'$

(5) $\beta_k \geq 0$ $(\beta_k = 0$ if $\bar{h}_k(t_1, \mathbf{x}^*(t_1)) > 0)$,

                    $k = s'+1, \ldots, s$

(6) $q_k(t) \geq 0$ $(= 0$ if $h_k(t, \mathbf{x}^*(t), \mathbf{u}^*(t)) > 0)$,

                    $k = 1, \ldots, s$

(7) $h_k(t, \mathbf{x}, \mathbf{u})$ is quasi-concave in $(\mathbf{x}, \mathbf{u})$,

                    $k = 1, \ldots, s$

(8) $H(t, \mathbf{x}, \mathbf{u}, \mathbf{p}(t))$ is concave in $(\mathbf{x}, \mathbf{u})$.

Then $(\mathbf{x}^*(t), \mathbf{u}^*(t))$ solves the problem.

Mangasarian's sufficient conditions for the mixed and pure state constraints problem (with $\mathbf{p}(t)$ continuous). $\mathcal{L}$ is defined in (16.53), and $\mathcal{L}^*$ denotes evaluation at $(t, \mathbf{x}^*(t), \mathbf{u}^*(t), \mathbf{p}(t), \mathbf{q}(t))$. $\mathbf{p}(t) = (p_1(t), \ldots, p_n(t))$, $\mathbf{q}(t) = (q_1(t), \ldots, q_s(t))$. A constraint qualification is not required, but the conditions often fail to hold because $\mathbf{p}(t)$ has discontinuities, in particular at t_1. See e.g. Seierstad and Sydsæter (1987), Theorem 6.2 for a sufficiency result allowing $\mathbf{p}(t)$ to have discontinuities at interior points of $[t_0, t_1]$ as well.

References

Kamien and Schwartz (1991), Léonard and Long (1992), Beavis and Dobbs (1990), Intriligator (1971), and Sydsæter et al. (2005). For more comprehensive collection of results, see e.g. Seierstad and Sydsæter (1987) or Feichtinger and Hartl (1986) (in German).

Chapter 17

Discrete dynamic optimization

Dynamic programming

17.1

$$\max \sum_{t=0}^{T} f(t, \mathbf{x}_t, \mathbf{u}_t)$$

$$\mathbf{x}_{t+1} = \mathbf{g}(t, \mathbf{x}_t, \mathbf{u}_t), \quad t = 0, \ldots, T-1$$

$$\mathbf{x}_0 = \mathbf{x}^0, \quad \mathbf{x}_t \in \mathbb{R}^n, \quad \mathbf{u}_t \in U \subset \mathbb{R}^r, \quad t = 0, \ldots, T$$

A *dynamic programming* problem. Here $\mathbf{g} = (g_1, \ldots, g_n)$, and $\mathbf{x}^0$ is a fixed vector in $\mathbb{R}^n$. U is the *control region*.

17.2

$$J_s(\mathbf{x}) = \max_{\mathbf{u}_s, \ldots, \mathbf{u}_T \in U} \sum_{t=s}^{T} f(t, \mathbf{x}_t, \mathbf{u}_t), \text{ where}$$

$$\mathbf{x}_{t+1} = \mathbf{g}(t, \mathbf{x}_t, \mathbf{u}_t), \quad t = s, \ldots, T-1, \quad \mathbf{x}_s = \mathbf{x}$$

Definition of the *value function*, $J_s(\mathbf{x})$, of problem (17.1).

17.3

$$J_T(\mathbf{x}) = \max_{\mathbf{u} \in U} f(T, \mathbf{x}, \mathbf{u})$$

$$J_s(\mathbf{x}) = \max_{\mathbf{u} \in U} \left[f(s, \mathbf{x}, \mathbf{u}) + J_{s+1}(\mathbf{g}(s, \mathbf{x}, \mathbf{u})) \right]$$

for $s = 0, 1, \ldots, T-1$.

The *fundamental equations* in dynamic programming. (Bellman's equations.)

A "control parameter free" formulation of the dynamic programming problem:

17.4

$$\max \sum_{t=0}^{T} F(t, \mathbf{x}_t, \mathbf{x}_{t+1})$$

$$\mathbf{x}_{t+1} \in \Gamma_t(\mathbf{x}_t), \quad t = 0, \ldots, T, \quad \mathbf{x}_0 \text{ given}$$

The set $\Gamma_t(\mathbf{x}_t)$ is often defined in terms of vector inequalities, $\mathbf{G}(t, \mathbf{x}_t) \le \mathbf{x}_{t+1} \le \mathbf{H}(t, \mathbf{x}_t)$, for given vector functions $\mathbf{G}$ and $\mathbf{H}$.

17.5

$$J_s(\mathbf{x}) = \max \sum_{t=s}^{T} F(t, \mathbf{x}_t, \mathbf{x}_{t+1}), \text{ where the maximum is taken over all } \mathbf{x}_{t+1} \text{ in } \Gamma_t(\mathbf{x}_t) \text{ for } t = s, \ldots, T, \text{ with } \mathbf{x}_s = \mathbf{x}.$$

The *value function*, $J_s(\mathbf{x})$, of problem (17.4).

17.6

$$J_T(\mathbf{x}) = \max_{\mathbf{y} \in \Gamma_T(\mathbf{x})} F(T, \mathbf{x}, \mathbf{y})$$

$$J_s(\mathbf{x}) = \max_{\mathbf{y} \in \Gamma_s(\mathbf{x})} \left[F(s, \mathbf{x}, \mathbf{y}) + J_{s+1}(\mathbf{y}) \right]$$

for $s = 0, 1, \ldots, T$.

The *fundamental equations* for problem (17.4).

17.7 If $\{\mathbf{x}_0^*, \ldots, \mathbf{x}_{T+1}^*\}$ is an optimal solution of problem (17.4) in which $\mathbf{x}_{t+1}^*$ is an interior point of $\Gamma_t(\mathbf{x}_t^*)$ for all t, and if the correspondence $\mathbf{x} \mapsto \mathbb{C}\Gamma_t(\mathbf{x})$ is upper hemicontinuous, then $\{\mathbf{x}_0^*, \ldots, \mathbf{x}_{T+1}^*\}$ satisfies the *Euler vector difference equation*

$$F_2'(t+1, \mathbf{x}_{t+1}, \mathbf{x}_{t+2}) + F_3'(t, \mathbf{x}_t, \mathbf{x}_{t+1}) = 0$$

F is a function of $1 + n + n$ variables, F_2' denotes the n-vector of partial derivatives of F w.r.t. variables no. 2, 3, ..., $n + 1$, and F_3' is the n-vector of partial derivatives of F w.r.t. variables no. $n + 2$, $n + 3$, ..., $2n + 1$.

Infinite horizon

17.8
$$\max \sum_{t=0}^{\infty} \alpha^t f(\mathbf{x}_t, \mathbf{u}_t)$$
$$\mathbf{x}_{t+1} = \mathbf{g}(\mathbf{x}_t, \mathbf{u}_t), \quad t = 0, 1, 2, \ldots$$
$$\mathbf{x}_0 = \mathbf{x}^0, \ \mathbf{x}_t \in \mathbb{R}^n, \ \mathbf{u}_t \in U \subset \mathbb{R}^r, \quad t = 0, 1, 2, \ldots$$

An infinite horizon problem. $\alpha \in (0, 1)$ is a constant discount factor.

17.9 The sequence $\{(\mathbf{x}_t, \mathbf{u}_t)\}$ is called *admissible* if $\mathbf{u}_t \in U$, $\mathbf{x}_0 = \mathbf{x}^0$, and the difference equation in (17.8) is satisfied for all $t = 0, 1, 2, \ldots$.

Definition of an admissible sequence.

17.10
(B) $\quad M \le f(\mathbf{x}, \mathbf{u}) \le N$
(BB) $\quad f(\mathbf{x}, \mathbf{u}) \ge M$
(BA) $\quad f(\mathbf{x}, \mathbf{u}) \le N$

Boundedness conditions. M and N are given numbers.

17.11 $V(\mathbf{x}, \boldsymbol{\pi}, s, \infty) = \sum_{t=s}^{\infty} \alpha^t f(\mathbf{x}_t, \mathbf{u}_t)$,
where $\boldsymbol{\pi} = (\mathbf{u}_s, \mathbf{u}_{s+1}, \ldots)$, with $\mathbf{u}_{s+k} \in U$ for $k = 0, 1, \ldots$, and with $\mathbf{x}_{t+1} = \mathbf{g}(\mathbf{x}_t, \mathbf{u}_t)$ for $t = s, s+1, \ldots$, and with $\mathbf{x}_s = \mathbf{x}$.

The total utility obtained from period s and onwards, given that the state vector is $\mathbf{x}$ at $t = s$.

17.12 $J_s(\mathbf{x}) = \sup_{\boldsymbol{\pi}} V(\mathbf{x}, \boldsymbol{\pi}, s, \infty)$
where the supremum is taken over all vectors $\boldsymbol{\pi} = (\mathbf{u}_s, \mathbf{u}_{s+1}, \ldots)$ with $\mathbf{u}_{s+k} \in U$, with $(\mathbf{x}_t, \mathbf{u}_t)$ admissible for $t \ge s$, and with $\mathbf{x}_s = \mathbf{x}$.

The value function of problem (17.8).

17.13
$$J_s(\mathbf{x}) = \alpha^s J_0(\mathbf{x}), \quad s = 1, 2, \ldots$$
$$J_0(\mathbf{x}) = \sup_{\mathbf{u} \in U} \{f(\mathbf{x}, \mathbf{u}) + \alpha J_0(\mathbf{g}(\mathbf{x}, \mathbf{u}))\}$$

Properties of the value function, assuming that at least one of the boundedness conditions in (17.10) is satisfied.

Discrete optimal control theory

17.14 $H = f(t, \mathbf{x}, \mathbf{u}) + \mathbf{p} \cdot \mathbf{g}(t, \mathbf{x}, \mathbf{u}), \quad t = 0, \ldots, T$

> The Hamiltonian $H = H(t, \mathbf{x}, \mathbf{u}, \mathbf{p})$ associated with (17.1), with $\mathbf{p} = (p^1, \ldots, p^n)$.

Suppose $\{(\mathbf{x}_t^*, \mathbf{u}_t^*)\}$ is an optimal sequence for problem (17.1). Then there exist vectors $\mathbf{p}_t$ in $\mathbb{R}^n$ such that for $t = 0, \ldots, T$:

17.15
  * $H_{\mathbf{u}}'(t, \mathbf{x}_t^*, \mathbf{u}_t^*, \mathbf{p}_t) \cdot (\mathbf{u} - \mathbf{u}_t^*) \le 0$ for all $\mathbf{u}$ in U
  * The vector $\mathbf{p}_t = (p_t^1, \ldots, p_t^n)$ is a solution of

$$\mathbf{p}_{t-1} = H_{\mathbf{x}}'(t, \mathbf{x}_t^*, \mathbf{u}_t^*, \mathbf{p}_t), \quad t = 1, \ldots, T$$

with $\mathbf{p}_T = \mathbf{0}$.

> The maximum principle for (17.1). Necessary conditions for optimality. U is convex. (The Hamiltonian is not necessarily maximized by $\mathbf{u}_t^*$.)

17.16
  (a) $x_T^i = \bar{x}^i \quad$ for $i = 1, \ldots, l$
  (b) $x_T^i \ge \bar{x}^i \quad$ for $i = l+1, \ldots, m$
  (c) x_T^i free $\quad$ for $i = m+1, \ldots, n$

> Terminal conditions for problem (17.1).

17.17 $H = \begin{cases} q_0 f(t, \mathbf{x}, \mathbf{u}) + \mathbf{p} \cdot \mathbf{g}(t, \mathbf{x}, \mathbf{u}), & t = 0, \ldots, T-1 \\ f(T, \mathbf{x}, \mathbf{u}), & t = T \end{cases}$

> The Hamiltonian $H = H(t, \mathbf{x}, \mathbf{u}, \mathbf{p})$ associated with (17.1) with terminal conditions (17.16).

Suppose $\{(\mathbf{x}_t^*, \mathbf{u}_t^*)\}$ is an optimal sequence for problem (17.1) with terminal conditions (17.16). Then there exist vectors $\mathbf{p}_t$ in $\mathbb{R}^n$ and a number q_0, with $(q_0, \mathbf{p}_T) \ne (0, \mathbf{0})$ and with $q_0 = 0$ or 1, such that for $t = 0, \ldots, T$:

17.18
  (1) $H_{\mathbf{u}}'(t, \mathbf{x}_t^*, \mathbf{u}_t^*, \mathbf{p}_t)(\mathbf{u} - \mathbf{u}_t^*) \le 0$ for all $\mathbf{u}$ in U
  (2) $\mathbf{p}_t = (p_t^1, \ldots, p_t^n)$ is a solution of

$$p_{t-1}^i = H_{x^i}'(t, \mathbf{x}_t^*, \mathbf{u}_t^*, \mathbf{p}_t), \quad t = 1, \ldots, T-1$$

  (3) $p_{T-1}^i = q_0 \dfrac{\partial f(T, \mathbf{x}_T^*, \mathbf{u}_T^*)}{\partial x_T^i} + p_T^i$

  where p_T^i satisfies

  (a') no condition on p_T^i, $\quad i = 1, \ldots, l$
  (b') $p_T^i \ge 0 \ (= 0$ if $x_T^{*i} > \bar{x}^i), \quad i = l+1, \ldots, m$
  (c') $p_T^i = 0, \quad i = m+1, \ldots, n$

> The maximum principle for (17.1) with terminal conditions (17.16). Necessary conditions for optimality. (a'), (b'), or (c') holds when (a), (b), or (c) in (17.16) holds, respectively. U is convex. (Except in degenerate cases, one can put $q_0 = 1$.)

Suppose that the sequence $\{(\mathbf{x}_t^*, \mathbf{u}_t^*, \mathbf{p}_t)\}$ satisfies all the conditions in (17.18) for $q_0 = 1$,
17.19 and suppose further that $H(t, \mathbf{x}, \mathbf{u}, \mathbf{p}_t)$ is concave in $(\mathbf{x}, \mathbf{u})$ for every $t \ge 0$. Then the sequence $\{(\mathbf{x}_t^*, \mathbf{u}_t^*, \mathbf{p}_t)\}$ is optimal.

> Sufficient conditions for optimality.

Infinite horizon

17.20
$$\max \sum_{t=0}^{\infty} f(t, \mathbf{x}_t, \mathbf{u}_t)$$
$$\mathbf{x}_{t+1} = \mathbf{g}(t, \mathbf{x}_t, \mathbf{u}_t), \quad t = 0, 1, 2, \ldots$$
$$\mathbf{x}_0 = \mathbf{x}^0, \quad \mathbf{x}_t \in \mathbb{R}^n, \quad \mathbf{u}_t \in U \subset \mathbb{R}^r, \quad t = 0, 1, 2, \ldots$$

It is assumed that the infinite sum converges for every admissible pair.

17.21
The sequence $\{(\mathbf{x}_t{}^*, \mathbf{u}_t{}^*)\}$ is *catching up optimal* (CU-optimal) if for every admissible sequence $\{(\mathbf{x}_t, \mathbf{u}_t)\}$,
$$\varliminf_{t \to \infty} D(t) \geq 0$$
where $D(t) = \sum_{\tau=0}^{t} (f(\tau, \mathbf{x}_\tau^*, \mathbf{u}_\tau^*) - f(\tau, \mathbf{x}_\tau, \mathbf{u}_\tau))$.

Definition of "catching up optimality". For $\varliminf$ see (12.42) and (12.43).

17.22
Suppose that the sequence $\{(\mathbf{x}_t^*, \mathbf{u}_t^*, \mathbf{p}_t)\}$ satisfies the conditions (1) and (2) in (17.18) with $q_0 = 1$. Suppose further that the Hamiltonian function $H(t, \mathbf{x}, \mathbf{u}, \mathbf{p}_t)$ is concave in $(\mathbf{x}, \mathbf{u})$ for every t. Then $\{(\mathbf{x}_t^*, \mathbf{u}_t^*)\}$ is CU-optimal provided that the following limit condition is satisfied: For all admissible sequences $\{(\mathbf{x}_t, \mathbf{u}_t)\}$,
$$\varliminf_{t \to \infty} \mathbf{p}_t \cdot (\mathbf{x}_t - \mathbf{x}_t^*) \geq 0$$

Sufficient optimality conditions for an infinite horizon problem with no terminal conditions.

References

See Bellman (1957), Stokey, Lucas, and Prescott (1989), and Sydsæter et al. (2005).

Chapter 18

Vectors in $\mathbb{R}^n$. Abstract spaces

18.1
$$\mathbf{a}_1 = \begin{pmatrix} a_{11} \\ a_{21} \\ \vdots \\ a_{n1} \end{pmatrix}, \ \mathbf{a}_2 = \begin{pmatrix} a_{12} \\ a_{22} \\ \vdots \\ a_{n2} \end{pmatrix}, \ \ldots, \ \mathbf{a}_m = \begin{pmatrix} a_{1m} \\ a_{2m} \\ \vdots \\ a_{nm} \end{pmatrix}$$

m (column) vectors in $\mathbb{R}^n$.

18.2 If $x_1, x_2, \ldots, x_m$ are real numbers, then
$$x_1\mathbf{a}_1 + x_2\mathbf{a}_2 + \cdots + x_m\mathbf{a}_m$$
is a *linear combination* of $\mathbf{a}_1, \mathbf{a}_2, \ldots, \mathbf{a}_m$.

Definition of a linear combination of vectors.

18.3 The vectors $\mathbf{a}_1, \mathbf{a}_2, \ldots, \mathbf{a}_m$ in $\mathbb{R}^n$ are

- *linearly dependent* if there exist numbers c_1, $c_2, \ldots, c_m$, not all zero, such that
$$c_1\mathbf{a}_1 + c_2\mathbf{a}_2 + \cdots + c_m\mathbf{a}_m = \mathbf{0}$$

- *linearly independent* if they are not linearly dependent.

Definition of linear dependence and independence.

18.4 The vectors $\mathbf{a}_1, \mathbf{a}_2, \ldots, \mathbf{a}_m$ in (18.1) are linearly independent if and only if the matrix $(a_{ij})_{n \times m}$ has rank m.

A characterization of linear independence for m vectors in $\mathbb{R}^n$. (See (19.23) for the definition of rank.)

18.5 The vectors $\mathbf{a}_1, \mathbf{a}_2, \ldots, \mathbf{a}_n$ in $\mathbb{R}^n$ are linearly independent if and only if
$$\begin{vmatrix} a_{11} & a_{12} & \cdots & a_{1n} \\ a_{21} & a_{22} & \cdots & a_{2n} \\ \vdots & \vdots & \ddots & \vdots \\ a_{n1} & a_{n2} & \cdots & a_{nn} \end{vmatrix} \neq 0$$

A characterization of linear independence for n vectors in $\mathbb{R}^n$. (A special case of (18.4).)

18.6 A non-empty subset V of vectors in $\mathbb{R}^n$ is a *subspace* of $\mathbb{R}^n$ if $c_1\mathbf{a}_1 + c_2\mathbf{a}_2 \in V$ for all $\mathbf{a}_1, \mathbf{a}_2$ in V and all numbers c_1, c_2.

Definition of a subspace.

18.7 If V is a subset of $\mathbb{R}^n$, then $\mathcal{S}[V]$ is the set of all linear combinations of vectors from V.

$\mathcal{S}[V]$ is called the *span* of V.

18.8 A collection of vectors $\mathbf{a}_1, \ldots, \mathbf{a}_m$ in a subspace V of $\mathbb{R}^n$ is a *basis* for V if the following two conditions are satisfied:

- $\mathbf{a}_1, \ldots, \mathbf{a}_m$ are linearly independent
- $\mathcal{S}[\mathbf{a}_1, \ldots, \mathbf{a}_m] = V$

Definition of a basis for a subspace.

18.9 The *dimension* $\dim V$, of a subspace V of $\mathbb{R}^n$ is the number of vectors in a basis for V. (Two bases for V always have the same number of vectors.)

Definition of the dimension of a subspace. In particular, $\dim \mathbb{R}^n = n$.

18.10 Let V be an m-dimensional subspace of $\mathbb{R}^n$.

- Any collection of m linearly independent vectors in V is a basis for V.

- Any collection of m vectors in V that spans V is a basis for V.

Important facts about subspaces.

18.11 The *inner product* of $\mathbf{a} = (a_1, \ldots, a_m)$ and $\mathbf{b} = (b_1, \ldots, b_m)$ is the number

$$\mathbf{a} \cdot \mathbf{b} = a_1 b_1 + \cdots + a_m b_m = \sum_{j=1}^{m} a_i b_i$$

Definition of the inner product, also called *scalar product* or *dot product*.

18.12
$$\mathbf{a} \cdot \mathbf{b} = \mathbf{b} \cdot \mathbf{a}$$
$$\mathbf{a} \cdot (\mathbf{b} + \mathbf{c}) = \mathbf{a} \cdot \mathbf{b} + \mathbf{a} \cdot \mathbf{c}$$
$$(\alpha \mathbf{a}) \cdot \mathbf{b} = \mathbf{a} \cdot (\alpha \mathbf{b}) = \alpha(\mathbf{a} \cdot \mathbf{b})$$
$$\mathbf{a} \cdot \mathbf{a} > 0 \iff \mathbf{a} \neq \mathbf{0}$$

Properties of the inner product. α is a scalar (i.e. a real number).

18.13 $\|\mathbf{a}\| = \sqrt{a_1^2 + a_2^2 + \cdots + a_n^2} = \sqrt{\mathbf{a} \cdot \mathbf{a}}$

Definition of the *(Euclidean) norm* (or *length*) of a vector.

18.14
(a) $\|\mathbf{a}\| > 0$ for $\mathbf{a} \neq \mathbf{0}$ and $\|\mathbf{0}\| = 0$
(b) $\|\alpha \mathbf{a}\| = |\alpha| \|\mathbf{a}\|$
(c) $\|\mathbf{a} + \mathbf{b}\| \leq \|\mathbf{a}\| + \|\mathbf{b}\|$
(d) $|\mathbf{a} \cdot \mathbf{b}| \leq \|\mathbf{a}\| \cdot \|\mathbf{b}\|$

Properties of the norm. $\mathbf{a}, \mathbf{b} \in \mathbb{R}^n$, α is a scalar. (d) is the *Cauchy–Schwarz inequality*. $\|\mathbf{a} - \mathbf{b}\|$ is the *distance* between $\mathbf{a}$ and $\mathbf{b}$.

18.15 The *angle* φ between two nonzero vectors $\mathbf{a}$ and $\mathbf{b}$ is defined by

$$\cos \varphi = \frac{\mathbf{a} \cdot \mathbf{b}}{\|\mathbf{a}\| \cdot \|\mathbf{b}\|}, \qquad 0 \leq \varphi \leq \pi$$

Definition of the angle between two vectors in $\mathbb{R}^n$. The vectors $\mathbf{a}$ and $\mathbf{b}$ are called *orthogonal* if $\mathbf{a} \cdot \mathbf{b} = 0$.

Vector spaces

A *vector space* (or *linear space*) (over $\mathbb{R}$) is a set V of elements, often called *vectors*, with two operations, "addition" $(V \times V \to V)$ and "scalar multiplication" $(\mathbb{R} \times V \to V)$, that for all x, y, z in V, and all real numbers α and β satisfy the following axioms:

18.16

   (a) $(x+y)+z = x+(y+z)$, $x+y = y+x$.

   (b) There is an element 0 in V with $x+0 = x$.

   (c) For every x in V, the element $(-1)x$ in V has the property $x + (-1)x = 0$.

   (d) $(\alpha+\beta)x = \alpha x + \beta x$, $\alpha(\beta x) = (\alpha\beta)x$, $\alpha(x+y) = \alpha x + \alpha y$, $1x = x$.

Definition of a vector space. With obvious modifications, definitions (18.2), (18.3), (18.6), and (18.7), of a linear combination, of linearly dependent and independent sets of vectors, of a subspace, and of the span, carry over to vector spaces.

18.17

A set B of vectors in a vector space V is a *basis* for V if the vectors in B are linearly independent, and B spans V, $\mathcal{S}[B] = V$.

Definition of a basis of a vector space.

Metric spaces

A *metric space* is a set M equipped with a *distance function* $d : M \times M \to \mathbb{R}$, such that the following axioms hold for all x, y, z in M:

18.18

   (a) $d(x,y) \geq 0$, and $d(x,y) = 0 \Leftrightarrow x = y$

   (b) $d(x,y) = d(y,x)$

   (c) $d(x,y) \leq d(x,z) + d(z,y)$

*Definition of a metric space. The distance function d is also called a *metric* on M. (c) is called the *triangle inequality*.*

A sequence $\{x_n\}$ in a metric space is

18.19

• *convergent* with limit x, and we write $\lim_{n\to\infty} x_n = x$ (or $x_n \to x$ as $n \to \infty$), if $d(x_n, x) \to 0$ as $n \to \infty$;

• a *Cauchy sequence* if for every $\varepsilon > 0$ there exists an integer N such that $d(x_n, x_m) < \varepsilon$ for all $m, n \geq N$.

*Important definitions. A sequence that is not convergent is called *divergent*.*

18.20

A subset S of a metric space M is *dense* in M if each point in M is the limit of a sequence of points in S.

Definition of a dense subset.

A metric space M is

18.21

• *complete* if every Cauchy sequence in M is convergent;

• *separable* if there exists a countable subset S of M that is dense in M.

Definition of complete and separable metric spaces.

Normed vector spaces. Banach spaces

18.22

A *normed vector space* (over $\mathbb{R}$) is a vector space V, together with a function $\| \cdot \| : V \to \mathbb{R}$, such that for all x, y in V and all real numbers α,

(a) $\|x\| > 0$ for $x \neq 0$ and $\|0\| = 0$

(b) $\|\alpha x\| = |\alpha| \, \|x\|$

(c) $\|x + y\| \leq \|x\| + \|y\|$

With the distance function $d(x, y) = \|x - y\|$, V becomes a metric space. If this metric space is complete, then V is called a *Banach space*.

18.23

- $l^p(n)$: $\mathbb{R}^n$, with $\|\mathbf{x}\| = \left(\sum_{i=1}^{n} |x_i|^p \right)^{1/p}$ $(p \geq 1)$
  (For $p = 2$ this is the Euclidean norm.)
- $l^\infty(n)$: $\mathbb{R}^n$, with $\|\mathbf{x}\| = \max(|x_1|, \ldots, |x_n|)$
- l^p $(p \geq 1)$: the set of all infinite sequences $\mathbf{x} = (x_0, x_1, \ldots)$ of real numbers such that $\sum_{i=1}^{\infty} |x_i|^p$ converges. $\|\mathbf{x}\| = \left(\sum_{i=1}^{\infty} |x_i|^p \right)^{1/p}$.
  For $\mathbf{x} = (x_0, x_1, \ldots)$ and $\mathbf{y} = (y_0, y_1, \ldots)$ in l^p, by definition, $\mathbf{x} + \mathbf{y} = (x_0 + y_0, x_1 + y_1, \ldots)$ and $\alpha \mathbf{x} = (\alpha x_0, \alpha x_1, \ldots)$.
- l^∞: the set of all *bounded* infinite sequences $\mathbf{x} = (x_0, x_1, \ldots)$ of real numbers, with $\|\mathbf{x}\| = \sup_i |x_i|$. (Vector operations defined as for l^p.)
- $C(X)$: the set of all bounded, continuous functions $f : X \to \mathbb{R}$, where X is a metric space, and with $\|f\| = \sup_{x \in X} |f(x)|$. If f and g are in $C(X)$ and $\alpha \in \mathbb{R}$, then $f + g$ and αf are defined by $(f + g)(x) = f(x) + g(x)$ and $(\alpha f)(x) = \alpha f(x)$.

Some standard examples of normed vector spaces, that are also Banach spaces.

18.24

Let X be compact metric space, and let F be a subset of the Banach space $C(X)$ (see (18.23)) that is

- *uniformly bounded*, i.e. there exists a number M such that $|f(x)| \leq M$ for all f in F and all x in X,
- *equicontinuous*, i.e. for each $\varepsilon > 0$ there exists a $\delta > 0$ such that if $\|x - x'\| < \delta$, then $|f(x) - f(x')| < \varepsilon$ for all f in F.

Then the closure of F is compact.

Ascoli's theorem. (Together with Schauder's theorem (18.25), this result is useful e.g. in economic dynamics. See Stokey, Lucas, and Prescott (1989).)

18.25 If K is a compact, convex set in a Banach space X, then any continuous function f of K into itself has a fixed point, i.e. there exists a point x^* in K such that $f(x^*) = x^*$.

Schauder's fixed point theorem.

18.26 Let $T : X \to X$ be a mapping of a complete metric space X into itself, and suppose there exists a number k in $[0, 1)$ such that

$(*)$ $d(Tx, Ty) \le kd(x, y)$ for all x, y in X

Then:

(a) T has a fixed point x^*, i.e. $T(x^*) = x^*$.

(b) $d(T^n x^0, x^*) \le k^n d(x^0, x^*)$ for all x^0 in X and all $n = 0, 1, 2, \ldots$.

The existence of a fixed point for a contraction mapping. k is called a *modulus* of the contraction mapping. (See also (6.23) and (6.25).) A mapping that satisfies $(*)$ for some k in $[0, 1)$, is called a *contraction mapping*.

18.27 Let $C(X)$ be the Banach space defined in (18.23) and let T be a mapping of $C(X)$ into $C(X)$ satisfying:

(a) (Monotonicity) If $f, g \in C(X)$ and $f(x) \le g(x)$ for all x in X, then $(Tf)(x) \le (Tg)(x)$ for all $x \in X$.

(b) (Discounting) There exists some α in $(0, 1)$ such that for all f in $C(X)$, all $a \ge 0$, and all x in X,

$[T(f + a)](x) \le (Tf)(x) + \alpha a$

Then T is a contraction mapping with modulus α.

Blackwell's sufficient conditions for a contraction. Here $(f + a)(x)$ is defined as $f(x) + a$.

Inner product spaces. Hilbert spaces

18.28 An *inner product space* (over $\mathbb{R}$) is a vector space V, together with a function that to each ordered pair of vectors (x, y) in V associates a real number, $\langle x, y \rangle$, such that for all x, y, z in V and all real numbers α,

(a) $\langle x, y \rangle = \langle y, x \rangle$

(b) $\langle x, y + z \rangle = \langle x, y \rangle + \langle x, z \rangle$

(c) $\alpha \langle x, y \rangle = \langle \alpha x, y \rangle = \langle x, \alpha y \rangle$

(d) $\langle x, x \rangle \ge 0$ and $\langle x, x \rangle = 0 \Leftrightarrow x = 0$

Definition of an inner product space. If we define $\|x\| = \sqrt{\langle x, x \rangle}$, then V becomes a normed vector space. If this space is complete, V is called a *Hilbert space*.

18.29

- $l^2(n)$, with $\langle \mathbf{x}, \mathbf{y} \rangle = \sum\limits_{i=1}^{n} x_i y_i$
- l^2, with $\langle \mathbf{x}, \mathbf{y} \rangle = \sum\limits_{i=1}^{\infty} x_i y_i$

Examples of Hilbert spaces.

18.30

(a) $|\langle x, y \rangle| \leq \sqrt{\langle x, x \rangle} \sqrt{\langle y, y \rangle}$ for all x, y in V

(b) $\langle x, y \rangle = \frac{1}{4}\big(\|x+y\|^2 - \|x-y\|^2\big)$

(a) is the Cauchy–Schwarz inequality. (Equality holds if and only if x and y are linearly dependent.) The equality in (b) shows that the inner product is expressible in terms of the norm.

18.31

- Two vectors x and y in an inner product space V are *orthogonal* if $\langle x, y \rangle = 0$.
- A set S of vectors in V is called *orthogonal* if $\langle x, y \rangle = 0$ for all $x \neq y$ in S.
- A set S of vectors in V is called *orthonormal* if it is orthogonal and $\|x\| = 1$ for all x in S.
- An orthonormal set S in V is called *complete* if there exists no x in V that is orthogonal to all vectors in S.

Important definitions.

18.32

Let U be an orthonormal set in an inner product space V.

(a) If $u_1, \ldots, u_n$ is any finite collection of distinct elements of U, then

$$(*) \quad \sum\limits_{i=1}^{n} |(x, u_i)|^2 \leq \|x\|^2 \quad \text{for all } x \text{ in } V$$

(b) If V is complete (a Hilbert space) and U is a complete orthonormal subset of V, then

$$(**) \quad \sum\limits_{u \in U} |(x, u)|^2 = \|x\|^2 \quad \text{for all } x \text{ in } V$$

$(*)$ is *Bessel's inequality*, $(**)$ is *Parseval's formula*.

References

All the results on vectors in $\mathbb{R}^n$ are standard and can be found in any linear algebra text, e.g. Fraleigh and Beauregard (1995) or Lang (1987). For abstract spaces, see Kolmogorov and Fomin (1975), or Royden (1968). For contraction mappings and their application in economic dynamics, see Stokey, Lucas, and Prescott (1989).

Chapter 19

Matrices

19.1 $\quad \mathbf{A} = \begin{pmatrix} a_{11} & a_{12} & \cdots & a_{1n} \\ a_{21} & a_{22} & \cdots & a_{2n} \\ \vdots & \vdots & & \vdots \\ a_{m1} & a_{m2} & \cdots & a_{mn} \end{pmatrix} = (a_{ij})_{m \times n}$

Notation for a *matrix*, where a_{ij} is the element in the ith row and the jth column. The matrix has *order* $m \times n$. If $m = n$, the matrix is *square* of order n.

19.2 $\quad \mathbf{A} = \begin{pmatrix} a_{11} & a_{12} & \cdots & a_{1n} \\ 0 & a_{22} & \cdots & a_{2n} \\ \vdots & \vdots & \ddots & \vdots \\ 0 & 0 & \cdots & a_{nn} \end{pmatrix}$

An *upper triangular* matrix. (All elements below the diagonal are 0.) The transpose of $\mathbf{A}$ (see (19.11)) is called *lower triangular*.

19.3 $\quad \mathrm{diag}(a_1, a_2, \ldots, a_n) = \begin{pmatrix} a_1 & 0 & \cdots & 0 \\ 0 & a_2 & \cdots & 0 \\ \vdots & \vdots & \ddots & \vdots \\ 0 & 0 & \cdots & a_n \end{pmatrix}$

A *diagonal matrix*.

19.4 $\quad \begin{pmatrix} a & 0 & \cdots & 0 \\ 0 & a & \cdots & 0 \\ \vdots & \vdots & \ddots & \vdots \\ 0 & 0 & \cdots & a \end{pmatrix}_{n \times n}$

A *scalar* matrix.

19.5 $\quad \mathbf{I}_n = \begin{pmatrix} 1 & 0 & \cdots & 0 \\ 0 & 1 & \cdots & 0 \\ \vdots & \vdots & \ddots & \vdots \\ 0 & 0 & \cdots & 1 \end{pmatrix}_{n \times n}$

The *unit* or *identity* matrix.

If $\mathbf{A} = (a_{ij})_{m \times n}$, $\mathbf{B} = (b_{ij})_{m \times n}$, and α is a scalar, we define

19.6 $\quad \mathbf{A} + \mathbf{B} = (a_{ij} + b_{ij})_{m \times n}$

$\qquad \alpha \mathbf{A} = (\alpha a_{ij})_{m \times n}$

$\qquad \mathbf{A} - \mathbf{B} = \mathbf{A} + (-1)\mathbf{B} = (a_{ij} - b_{ij})_{m \times n}$

Matrix operations. (The scalars are real or complex numbers.)

$$\begin{aligned}
(\mathbf{A} + \mathbf{B}) + \mathbf{C} &= \mathbf{A} + (\mathbf{B} + \mathbf{C})\\
\mathbf{A} + \mathbf{B} &= \mathbf{B} + \mathbf{A}\\
\mathbf{A} + \mathbf{0} &= \mathbf{A}\\
\mathbf{A} + (-\mathbf{A}) &= \mathbf{0}\\
(a + b)\mathbf{A} &= a\mathbf{A} + b\mathbf{A}\\
a(\mathbf{A} + \mathbf{B}) &= a\mathbf{A} + a\mathbf{B}
\end{aligned}$$

19.7

Properties of matrix operations. $\mathbf{0}$ is the zero (or null) matrix, all of whose elements are zero. a and b are scalars.

19.8 If $\mathbf{A} = (a_{ij})_{m \times n}$ and $\mathbf{B} = (b_{jk})_{n \times p}$, we define the *product* $\mathbf{C} = \mathbf{AB}$ as the $m \times p$ matrix $\mathbf{C} = (c_{ik})_{m \times p}$ where

$$c_{ik} = a_{i1}b_{1k} + \cdots + a_{ij}b_{jk} + \cdots + a_{in}b_{nk}$$

The definition of *matrix multiplication*.

$$
\begin{pmatrix}
a_{11} & \cdots & a_{1j} & \cdots & a_{1n}\\
\vdots & & \vdots & & \vdots\\
a_{i1} & \cdots & a_{ij} & \cdots & a_{in}\\
\vdots & & \vdots & & \vdots\\
a_{m1} & \cdots & a_{mj} & \cdots & a_{mn}
\end{pmatrix}
\cdot
\begin{pmatrix}
b_{11} & \cdots & b_{1k} & \cdots & b_{1p}\\
\vdots & & \vdots & & \vdots\\
b_{j1} & \cdots & b_{jk} & \cdots & b_{jp}\\
\vdots & & \vdots & & \vdots\\
b_{n1} & \cdots & b_{nk} & \cdots & b_{np}
\end{pmatrix}
=
\begin{pmatrix}
c_{11} & \cdots & c_{1k} & \cdots & c_{1p}\\
\vdots & & \vdots & & \vdots\\
c_{i1} & \cdots & c_{ik} & \cdots & c_{ip}\\
\vdots & & \vdots & & \vdots\\
c_{m1} & \cdots & c_{mk} & \cdots & c_{mp}
\end{pmatrix}
$$

19.9
$$\begin{aligned}
(\mathbf{AB})\mathbf{C} &= \mathbf{A}(\mathbf{BC})\\
\mathbf{A}(\mathbf{B} + \mathbf{C}) &= \mathbf{AB} + \mathbf{AC}\\
(\mathbf{A} + \mathbf{B})\mathbf{C} &= \mathbf{AC} + \mathbf{BC}
\end{aligned}$$

Properties of matrix multiplication.

19.10
$$\begin{aligned}
\mathbf{AB} &\neq \mathbf{BA}\\
\mathbf{AB} = \mathbf{0} &\not\Rightarrow \mathbf{A} = \mathbf{0} \text{ or } \mathbf{B} = \mathbf{0}\\
\mathbf{AB} = \mathbf{AC} \ \& \ \mathbf{A} \neq \mathbf{0} &\not\Rightarrow \mathbf{B} = \mathbf{C}
\end{aligned}$$

Important observations about matrix multiplication. $\mathbf{0}$ is the zero matrix. $\not\Rightarrow$ should be read: "does not necessarily imply".

19.11
$$\mathbf{A}' = \begin{pmatrix}
a_{11} & a_{21} & \cdots & a_{m1}\\
a_{12} & a_{22} & \cdots & a_{m2}\\
\vdots & \vdots & & \vdots\\
a_{1n} & a_{2n} & \cdots & a_{mn}
\end{pmatrix}$$

$\mathbf{A}'$, the *transpose* of $\mathbf{A} = (a_{ij})_{m \times n}$, is the $n \times m$ matrix obtained by interchanging rows and columns in $\mathbf{A}$.

19.12
$$\begin{aligned}
(\mathbf{A}')' &= \mathbf{A}\\
(\mathbf{A} + \mathbf{B})' &= \mathbf{A}' + \mathbf{B}'\\
(\alpha\mathbf{A})' &= \alpha\mathbf{A}'\\
(\mathbf{AB})' &= \mathbf{B}'\mathbf{A}' \quad \text{(NOTE the order!)}
\end{aligned}$$

Rules for transposes.

19.13 $\mathbf{B} = \mathbf{A}^{-1} \iff \mathbf{AB} = \mathbf{I}_n \iff \mathbf{BA} = \mathbf{I}_n$

The *inverse* of an $n \times n$ matrix $\mathbf{A}$. $\mathbf{I}_n$ is the identity matrix.

19.14 $\quad \mathbf{A}^{-1}$ exists $\iff |\mathbf{A}| \neq 0$

A necessary and sufficient condition for a matrix to have an inverse, i.e. to be *invertible*. $|\mathbf{A}|$ denotes the determinant of the square matrix $\mathbf{A}$. (See Chapter 20.)

19.15 $\quad \mathbf{A} = \begin{pmatrix} a & b \\ c & d \end{pmatrix} \implies \mathbf{A}^{-1} = \dfrac{1}{ad-bc} \begin{pmatrix} d & -b \\ -c & a \end{pmatrix}$

Valid if
$|\mathbf{A}| = ad - bc \neq 0$.

19.16

If $\mathbf{A} = (a_{ij})_{n \times n}$ is a square matrix and $|\mathbf{A}| \neq 0$, the unique inverse of $\mathbf{A}$ is given by

$$\mathbf{A}^{-1} = \frac{1}{|\mathbf{A}|} \operatorname{adj}(\mathbf{A}), \quad \text{where}$$

$$\operatorname{adj}(\mathbf{A}) = \begin{pmatrix} A_{11} & A_{21} & \cdots & A_{n1} \\ A_{12} & A_{22} & \cdots & A_{n2} \\ \vdots & \vdots & \ddots & \vdots \\ A_{1n} & A_{2n} & \cdots & A_{nn} \end{pmatrix}$$

Here the *cofactor*, A_{ij}, of the element a_{ij} is given by

$$A_{ij} = (-1)^{i+j} \begin{vmatrix} a_{11} & \cdots & a_{1j} & \cdots & a_{1n} \\ \vdots & & \vdots & & \vdots \\ a_{i1} & \cdots & a_{ij} & \cdots & a_{in} \\ \vdots & & \vdots & & \vdots \\ a_{n1} & \cdots & a_{nj} & \cdots & a_{nn} \end{vmatrix}$$

The general formula for the inverse of a square matrix. NOTE the order of the indices in the *adjoint matrix*, $\operatorname{adj}(\mathbf{A})$. The matrix $(A_{ij})_{n \times n}$ is called the *cofactor matrix*, and thus the adjoint is the transpose of the cofactor matrix. In the formula for the cofactor, A_{ij}, the determinant is obtained by deleting the ith row and the jth column in $|\mathbf{A}|$.

19.17

$(\mathbf{A}^{-1})^{-1} = \mathbf{A}$
$(\mathbf{AB})^{-1} = \mathbf{B}^{-1}\mathbf{A}^{-1}$ (NOTE the order!)
$(\mathbf{A}')^{-1} = (\mathbf{A}^{-1})'$
$(c\mathbf{A})^{-1} = c^{-1}\mathbf{A}^{-1}$

Properties of the inverse. ($\mathbf{A}$ and $\mathbf{B}$ are invertible $n \times n$ matrices. c is a scalar $\neq 0$.)

19.18 $\quad (\mathbf{I}_m + \mathbf{AB})^{-1} = \mathbf{I}_m - \mathbf{A}(\mathbf{I}_n + \mathbf{BA})^{-1}\mathbf{B}$

$\mathbf{A}$ is $m \times n$, $\mathbf{B}$ is $n \times m$, $|\mathbf{I}_m + \mathbf{AB}| \neq 0$.

19.19 $\quad \mathbf{R}^{-1}\mathbf{A}'(\mathbf{A}\mathbf{R}^{-1}\mathbf{A}'+\mathbf{Q}^{-1})^{-1} = (\mathbf{A}'\mathbf{Q}\mathbf{A}+\mathbf{R})^{-1}\mathbf{A}'\mathbf{Q}$

Matrix inversion pairs. Valid if the inverses exist.

A square matrix $\mathbf{A}$ of order n is called

- *symmetric* if $\mathbf{A} = \mathbf{A}'$
- *skew-symmetric* if $\mathbf{A} = -\mathbf{A}'$

19.20
- *idempotent* if $\mathbf{A}^2 = \mathbf{A}$
- *involutive* if $\mathbf{A}^2 = \mathbf{I}_n$
- *orthogonal* if $\mathbf{A}'\mathbf{A} = \mathbf{I}_n$
- *singular* if $|\mathbf{A}| = 0$, *nonsingular* if $|\mathbf{A}| \neq 0$

Some important definitions. $|\mathbf{A}|$ denotes the determinant of the square matrix $\mathbf{A}$. (See Chapter 20.) For properties of idempotent and orthogonal matrices, see Chapter 22.

19.21 $\quad \text{tr}(\mathbf{A}) = \sum_{i=1}^{n} a_{ii}$

The *trace* of $\mathbf{A} = (a_{ij})_{n \times n}$ is the sum of its diagonal elements.

19.22
$$\text{tr}(\mathbf{A} + \mathbf{B}) = \text{tr}(\mathbf{A}) + \text{tr}(\mathbf{B})$$
$$\text{tr}(c\mathbf{A}) = c\,\text{tr}(\mathbf{A}) \quad (c \text{ is a scalar})$$
$$\text{tr}(\mathbf{AB}) = \text{tr}(\mathbf{BA}) \quad (\text{if } \mathbf{AB} \text{ is a square matrix})$$
$$\text{tr}(\mathbf{A}') = \text{tr}(\mathbf{A})$$

Properties of the trace.

19.23 $\quad r(\mathbf{A})$ = maximum number of linearly independent rows in $\mathbf{A}$ = maximum number of linearly independent columns in $\mathbf{A}$ = order of the largest nonzero minor of $\mathbf{A}$.

Equivalent definitions of the *rank* of a matrix. On minors, see (20.15).

19.24
(1) $r(\mathbf{A}) = r(\mathbf{A}') = r(\mathbf{A}'\mathbf{A}) = r(\mathbf{AA}')$
(2) $r(\mathbf{AB}) \leq \min(r(\mathbf{A}), r(\mathbf{B}))$
(3) $r(\mathbf{AB}) = r(\mathbf{B})$ if $|\mathbf{A}| \neq 0$
(4) $r(\mathbf{CA}) = r(\mathbf{C})$ if $|\mathbf{A}| \neq 0$
(5) $r(\mathbf{PAQ}) = r(\mathbf{A})$ if $|\mathbf{P}| \neq 0$, $|\mathbf{Q}| \neq 0$
(6) $|r(\mathbf{A}) - r(\mathbf{B})| \leq r(\mathbf{A} + \mathbf{B}) \leq r(\mathbf{A}) + r(\mathbf{B})$
(7) $r(\mathbf{AB}) \geq r(\mathbf{A}) + r(\mathbf{B}) - n$
(8) $r(\mathbf{AB}) + r(\mathbf{BC}) \leq r(\mathbf{B}) + r(\mathbf{ABC})$

Properties of the rank. The orders of the matrices are such that the required operations are defined. In result (7), *Sylvester's inequality*, $\mathbf{A}$ is $m \times n$ and $\mathbf{B}$ is $n \times p$. (8) is called *Frobenius's inequality*.

19.25 $\quad \mathbf{Ax} = \mathbf{0}$ for some $\mathbf{x} \neq \mathbf{0} \iff r(\mathbf{A}) \leq n - 1$

A useful result on homogeneous equations. $\mathbf{A}$ is $m \times n$, $\mathbf{x}$ is $n \times 1$.

A matrix norm is a function $\| \cdot \|_\beta$ that to each square matrix $\mathbf{A}$ associates a real number $\|\mathbf{A}\|_\beta$ such that:

19.26
- $\|\mathbf{A}\|_\beta > 0$ for $\mathbf{A} \neq \mathbf{0}$ and $\|\mathbf{0}\|_\beta = 0$
- $\|c\mathbf{A}\|_\beta = |c|\,\|\mathbf{A}\|_\beta \quad (c \text{ is a scalar})$
- $\|\mathbf{A} + \mathbf{B}\|_\beta \leq \|\mathbf{A}\|_\beta + \|\mathbf{B}\|_\beta$
- $\|\mathbf{AB}\|_\beta \leq \|\mathbf{A}\|_\beta \|\mathbf{B}\|_\beta$

Definition of a *matrix norm*. (There are an infinite number of such norms, some of which are given in (19.27).)

- $\|\mathbf{A}\|_1 = \max_{j=1,\dots,n} \sum_{i=1}^{n} |a_{ij}|$

- $\|\mathbf{A}\|_\infty = \max_{i=1,\dots,n} \sum_{j=1}^{n} |a_{ij}|$

19.27
- $\|\mathbf{A}\|_2 = \sqrt{\lambda}$, where λ is the largest eigenvalue of $\mathbf{A}'\mathbf{A}$.

- $\|\mathbf{A}\|_M = n \max_{i,j=1,\dots,n} |a_{ij}|$

- $\|\mathbf{A}\|_T = \left(\sum_{j=1}^{n} \sum_{i=1}^{n} |a_{ij}|^2 \right)^{1/2}$

Some matrix norms for $\mathbf{A} = (a_{ij})_{n \times n}$. (For eigenvalues, see Chapter 21.)

19.28 λ eigenvalue of $\mathbf{A} = (a_{ij})_{n \times n} \Rightarrow |\lambda| \le \|\mathbf{A}\|_\beta$

The modulus of any eigenvalue of $\mathbf{A}$ is less than or equal to any matrix norm of $\mathbf{A}$.

19.29 $\|\mathbf{A}\|_\beta < 1 \Rightarrow \mathbf{A}^t \to \mathbf{0}$ as $t \to \infty$

Sufficient condition for $\mathbf{A}^t \to \mathbf{0}$ as $t \to \infty$. $\|\mathbf{A}\|_\beta$ is any matrix norm of $\mathbf{A}$.

19.30 $e^{\mathbf{A}} = \sum_{n=0}^{\infty} \frac{1}{n!} \mathbf{A}^n$

The *exponential matrix* of a square matrix $\mathbf{A}$.

19.31
$e^{\mathbf{A}+\mathbf{B}} = e^{\mathbf{A}} e^{\mathbf{B}}$ if $\mathbf{AB} = \mathbf{BA}$

$(e^{\mathbf{A}})^{-1} = e^{-\mathbf{A}}, \quad \frac{d}{dx}(e^{x\mathbf{A}}) = \mathbf{A}e^{x\mathbf{A}}$

Properties of the exponential matrix.

Linear transformations

19.32
A function $T : \mathbb{R}^n \to \mathbb{R}^m$ is called a *linear* transformation (or function) if

(1) $T(\mathbf{x} + \mathbf{y}) = T(\mathbf{x}) + T(\mathbf{y})$

(2) $T(c\mathbf{x}) = cT(\mathbf{x})$

for all $\mathbf{x}$ and $\mathbf{y}$ in $\mathbb{R}^n$ and for all scalars c.

Definition of a linear transformation.

19.33
If $\mathbf{A}$ is an $m \times n$ matrix, the function $T_{\mathbf{A}} : \mathbb{R}^n \to \mathbb{R}^m$ defined by $T_{\mathbf{A}}(\mathbf{x}) = \mathbf{A}\mathbf{x}$ is a linear transformation.

An important fact.

19.34
Let $T : \mathbb{R}^n \to \mathbb{R}^m$ be a linear transformation and let $\mathbf{A}$ be the $m \times n$ matrix whose jth column is $T(\mathbf{e}_j)$, where $\mathbf{e}_j$ is the jth standard unit vector in $\mathbb{R}^n$. Then $T(\mathbf{x}) = \mathbf{A}\mathbf{x}$ for all $\mathbf{x}$ in $\mathbb{R}^n$.

The matrix $\mathbf{A}$ is called the *standard matrix representation* of T.

19.35 Let $T : \mathbb{R}^n \to \mathbb{R}^m$ and $S : \mathbb{R}^m \to \mathbb{R}^k$ be two linear transformations with standard matrix representations $\mathbf{A}$ and $\mathbf{B}$, respectively. Then the composition $S \circ T$ of the two transformations is a linear transformation with standard matrix representation $\mathbf{BA}$.

A basic fact.

19.36 Let $\mathbf{A}$ be an invertible $n \times n$ matrix with associated linear transformation T. The transformation T^{-1} associated with $\mathbf{A}^{-1}$ is the inverse transformation (function) of T.

A basic fact.

Generalized inverses

19.37 An $n \times m$ matrix $\mathbf{A}^-$ is called a *generalized inverse* of the $m \times n$ matrix $\mathbf{A}$ if

$$\mathbf{A}\mathbf{A}^-\mathbf{A} = \mathbf{A}$$

Definition of a generalized inverse of a matrix. ($\mathbf{A}^-$ is not unique in general.)

19.38 A necessary and sufficient condition for the matrix equation $\mathbf{A}\mathbf{x} = \mathbf{b}$ to have a solution is that $\mathbf{A}\mathbf{A}^-\mathbf{b} = \mathbf{b}$. The general solution is then $\mathbf{x} = \mathbf{A}^-\mathbf{b} + (\mathbf{I} - \mathbf{A}^-\mathbf{A})\mathbf{q}$, where $\mathbf{q}$ is an arbitrary vector of appropriate order.

An important application of generalized inverses.

19.39 If $\mathbf{A}^-$ is a generalized inverse of $\mathbf{A}$, then

- $\mathbf{A}\mathbf{A}^-$ and $\mathbf{A}^-\mathbf{A}$ are idempotent
- $r(\mathbf{A}) = r(\mathbf{A}^-\mathbf{A}) = \mathrm{tr}(\mathbf{A}^-\mathbf{A})$
- $(\mathbf{A}^-)'$ is a generalized inverse of $\mathbf{A}'$
- $\mathbf{A}$ is square and nonsingular $\Rightarrow \mathbf{A}^- = \mathbf{A}^{-1}$

Properties of generalized inverses.

19.40 An $n \times m$ matrix $\mathbf{A}^+$ is called the *Moore–Penrose inverse* of a real $m \times n$ matrix $\mathbf{A}$ if it satisfies the following four conditions:

(i) $\mathbf{A}\mathbf{A}^+\mathbf{A} = \mathbf{A}$ (ii) $\mathbf{A}^+\mathbf{A}\mathbf{A}^+ = \mathbf{A}^+$

(iii) $(\mathbf{A}\mathbf{A}^+)' = \mathbf{A}\mathbf{A}^+$ (iv) $(\mathbf{A}^+\mathbf{A})' = \mathbf{A}^+\mathbf{A}$

Definition of the Moore–Penrose inverse. ($\mathbf{A}^+$ exists and is unique.)

19.41 A necessary and sufficient condition for the matrix equation $\mathbf{A}\mathbf{x} = \mathbf{b}$ to have a solution is that $\mathbf{A}\mathbf{A}^+\mathbf{b} = \mathbf{b}$. The general solution is then $\mathbf{x} = \mathbf{A}^+\mathbf{b} + (\mathbf{I} - \mathbf{A}^+\mathbf{A})\mathbf{q}$, where $\mathbf{q}$ is an arbitrary vector of appropriate order.

An important application of the Moore–Penrose inverse.

- $\mathbf{A}$ is square and nonsingular $\Rightarrow \mathbf{A}^+ = \mathbf{A}^{-1}$
- $(\mathbf{A}^+)^+ = \mathbf{A}$, $(\mathbf{A}')^+ = (\mathbf{A}^+)'$
- $\mathbf{A}^+ = \mathbf{A}$ if $\mathbf{A}$ is symmetric and idempotent.
- $\mathbf{A}^+\mathbf{A}$ and $\mathbf{A}\mathbf{A}^+$ are idempotent.
- 19.42 $\mathbf{A}$, $\mathbf{A}^+$, $\mathbf{A}\mathbf{A}^+$, and $\mathbf{A}^+\mathbf{A}$ have the same rank.
- $\mathbf{A}'\mathbf{A}\mathbf{A}^+ = \mathbf{A}' = \mathbf{A}^+\mathbf{A}\mathbf{A}'$
- $(\mathbf{A}\mathbf{A}^+)^+ = \mathbf{A}\mathbf{A}^+$
- $(\mathbf{A}'\mathbf{A})^+ = \mathbf{A}^+(\mathbf{A}^+)'$, $(\mathbf{A}\mathbf{A}')^+ = (\mathbf{A}^+)'\mathbf{A}^+$
- $(\mathbf{A} \otimes \mathbf{B})^+ = \mathbf{A}^+ \otimes \mathbf{B}^+$

Properties of the Moore–Penrose inverse. ($\otimes$ is the Kronecker product. See Chapter 23.)

Partitioned matrices

19.43 $\mathbf{P} = \begin{pmatrix} \mathbf{P}_{11} & \mathbf{P}_{12} \\ \mathbf{P}_{21} & \mathbf{P}_{22} \end{pmatrix}$

A *partitioned* matrix of order $(p + q) \times (r + s)$. ($\mathbf{P}_{11}$ is $p \times r$, $\mathbf{P}_{12}$ is $p \times s$, $\mathbf{P}_{21}$ is $q \times r$, $\mathbf{P}_{22}$ is $q \times s$.)

19.44 $\begin{pmatrix} \mathbf{P}_{11} & \mathbf{P}_{12} \\ \mathbf{P}_{21} & \mathbf{P}_{22} \end{pmatrix} \begin{pmatrix} \mathbf{Q}_{11} & \mathbf{Q}_{12} \\ \mathbf{Q}_{21} & \mathbf{Q}_{22} \end{pmatrix}$

$= \begin{pmatrix} \mathbf{P}_{11}\mathbf{Q}_{11} + \mathbf{P}_{12}\mathbf{Q}_{21} & \mathbf{P}_{11}\mathbf{Q}_{12} + \mathbf{P}_{12}\mathbf{Q}_{22} \\ \mathbf{P}_{21}\mathbf{Q}_{11} + \mathbf{P}_{22}\mathbf{Q}_{21} & \mathbf{P}_{21}\mathbf{Q}_{12} + \mathbf{P}_{22}\mathbf{Q}_{22} \end{pmatrix}$

Multiplication of partitioned matrices. (We assume that the multiplications involved are defined.)

19.45 $\begin{vmatrix} \mathbf{P}_{11} & \mathbf{P}_{12} \\ \mathbf{P}_{21} & \mathbf{P}_{22} \end{vmatrix} = |\mathbf{P}_{11}| \cdot |\mathbf{P}_{22} - \mathbf{P}_{21}\mathbf{P}_{11}^{-1}\mathbf{P}_{12}|$

The determinant of a partitioned $n \times n$ matrix, assuming $\mathbf{P}_{11}^{-1}$ exists.

19.46 $\begin{vmatrix} \mathbf{P}_{11} & \mathbf{P}_{12} \\ \mathbf{P}_{21} & \mathbf{P}_{22} \end{vmatrix} = |\mathbf{P}_{22}| \cdot |\mathbf{P}_{11} - \mathbf{P}_{12}\mathbf{P}_{22}^{-1}\mathbf{P}_{21}|$

The determinant of a partitioned $n \times n$ matrix, assuming $\mathbf{P}_{22}^{-1}$ exists.

19.47 $\begin{vmatrix} \mathbf{P}_{11} & \mathbf{P}_{12} \\ \mathbf{0} & \mathbf{P}_{22} \end{vmatrix} = |\mathbf{P}_{11}| \cdot |\mathbf{P}_{22}|$

A special case.

19.48 $\begin{pmatrix} \mathbf{P}_{11} & \mathbf{P}_{12} \\ \mathbf{P}_{21} & \mathbf{P}_{22} \end{pmatrix}^{-1} =$

$\begin{pmatrix} \mathbf{P}_{11}^{-1} + \mathbf{P}_{11}^{-1}\mathbf{P}_{12}\boldsymbol{\Delta}^{-1}\mathbf{P}_{21}\mathbf{P}_{11}^{-1} & -\mathbf{P}_{11}^{-1}\mathbf{P}_{12}\boldsymbol{\Delta}^{-1} \\ -\boldsymbol{\Delta}^{-1}\mathbf{P}_{21}\mathbf{P}_{11}^{-1} & \boldsymbol{\Delta}^{-1} \end{pmatrix}$

where $\boldsymbol{\Delta} = \mathbf{P}_{22} - \mathbf{P}_{21}\mathbf{P}_{11}^{-1}\mathbf{P}_{12}$.

The inverse of a partitioned matrix, assuming $\mathbf{P}_{11}^{-1}$ exists.

$$\begin{pmatrix} \mathbf{P}_{11} & \mathbf{P}_{12} \\ \mathbf{P}_{21} & \mathbf{P}_{22} \end{pmatrix}^{-1} =$$

19.49
$$\begin{pmatrix} \boldsymbol{\Delta}_1^{-1} & -\boldsymbol{\Delta}_1^{-1}\mathbf{P}_{12}\mathbf{P}_{22}^{-1} \\ -\mathbf{P}_{22}^{-1}\mathbf{P}_{21}\boldsymbol{\Delta}_1^{-1} & \mathbf{P}_{22}^{-1} + \mathbf{P}_{22}^{-1}\mathbf{P}_{21}\boldsymbol{\Delta}_1^{-1}\mathbf{P}_{12}\mathbf{P}_{22}^{-1} \end{pmatrix}$$

where $\boldsymbol{\Delta}_1 = \mathbf{P}_{11} - \mathbf{P}_{12}\mathbf{P}_{22}^{-1}\mathbf{P}_{21}$.

The inverse of a partitioned matrix, assuming $\mathbf{P}_{22}^{-1}$ exists.

Matrices with complex elements

Let $\mathbf{A} = (a_{ij})$ be a complex matrix (i.e. the elements of $\mathbf{A}$ are complex numbers). Then:

19.50
- $\bar{\mathbf{A}} = (\bar{a}_{ij})$ is called the *conjugate* of $\mathbf{A}$. ($\bar{a}_{ij}$ denotes the complex conjugate of a_{ij}.)
- $\mathbf{A}^* = \bar{\mathbf{A}}' = (\bar{a}_{ji})$ is called the *conjugate transpose* of $\mathbf{A}$.
- $\mathbf{A}$ is called *Hermitian* if $\mathbf{A} = \mathbf{A}^*$.
- $\mathbf{A}$ is called *unitary* if $\mathbf{A}^* = \mathbf{A}^{-1}$.

Useful definitions in connection with complex matrices.

19.51
- $\mathbf{A}$ is real $\iff \mathbf{A} = \bar{\mathbf{A}}$.
- If $\mathbf{A}$ is real, then
  $$\mathbf{A} \text{ is Hermitian} \iff \mathbf{A} \text{ is symmetric}.$$

Easy consequences of the definitions.

Let $\mathbf{A}$ and $\mathbf{B}$ be complex matrices and c a complex number. Then

19.52
(1) $(\mathbf{A}^*)^* = \mathbf{A}$

(2) $(\mathbf{A} + \mathbf{B})^* = \mathbf{A}^* + \mathbf{B}^*$

(3) $(c\mathbf{A})^* = \bar{c}\mathbf{A}^*$

(4) $(\mathbf{AB})^* = \mathbf{B}^*\mathbf{A}^*$

Properties of the conjugate transpose. (2) and (4) are valid if the sum and the product of the matrices are defined.

References

Most of the formulas are standard and can be found in almost any linear algebra text, e.g. Fraleigh and Beauregard (1995) or Lang (1987). See also Sydsæter and Hammond (2005) and Sydsæter et al. (2005). For (19.26)–(19.29), see e.g. Faddeeva (1959). For generalized inverses, see Magnus and Neudecker (1988). A standard reference is Gantmacher (1959).

Chapter 20

Determinants

20.1
$$\begin{vmatrix} a_{11} & a_{12} \\ a_{21} & a_{22} \end{vmatrix} = a_{11}a_{22} - a_{21}a_{12}$$

Definition of a 2 × 2 determinant.

20.2

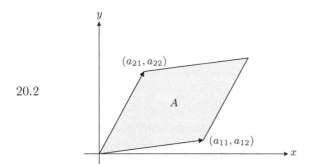

Geometric interpretation of a 2 × 2 determinant. The area A is the absolute value of the determinant
$$\begin{vmatrix} a_{11} & a_{12} \\ a_{21} & a_{22} \end{vmatrix}.$$

20.3
$$\begin{vmatrix} a_{11} & a_{12} & a_{13} \\ a_{21} & a_{22} & a_{23} \\ a_{31} & a_{32} & a_{33} \end{vmatrix} = \begin{cases} a_{11}a_{22}a_{33} - a_{11}a_{23}a_{32} \\ + a_{12}a_{23}a_{31} - a_{12}a_{21}a_{33} \\ + a_{13}a_{21}a_{32} - a_{13}a_{22}a_{31} \end{cases}$$

Definition of a 3 × 3 determinant.

20.4

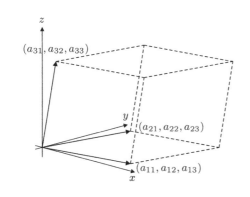

Geometric interpretation of a 3 × 3 determinant. The volume of the "box" spanned by the three vectors is the absolute value of the determinant
$$\begin{vmatrix} a_{11} & a_{12} & a_{13} \\ a_{21} & a_{22} & a_{23} \\ a_{31} & a_{32} & a_{33} \end{vmatrix}.$$

If $\mathbf{A} = (a_{ij})_{n \times n}$ is an $n \times n$ matrix, the *determinant* of $\mathbf{A}$ is the number

$$|\mathbf{A}| = a_{i1}A_{i1} + \cdots + a_{in}A_{in} = \sum_{j=1}^{n} a_{ij}A_{ij}$$

where A_{ij}, the cofactor of the element a_{ij}, is

20.5

$$A_{ij} = (-1)^{i+j} \begin{vmatrix} a_{11} & \cdots & a_{1j} & \cdots & a_{1n} \\ \vdots & & & & \vdots \\ a_{i1} & \cdots & \boxed{a_{ij}} & \cdots & a_{in} \\ \vdots & & & & \vdots \\ a_{n1} & \cdots & a_{nj} & \cdots & a_{nn} \end{vmatrix}$$

The general definition of a determinant of order n, by cofactor expansion along the ith row. The value of the determinant is independent of the choice of i.

20.6

$$a_{i1}A_{i1} + a_{i2}A_{i2} + \cdots + a_{in}A_{in} = |\mathbf{A}|$$
$$a_{i1}A_{k1} + a_{i2}A_{k2} + \cdots + a_{in}A_{kn} = 0 \quad \text{if } k \neq i$$

$$a_{1j}A_{1j} + a_{2j}A_{2j} + \cdots + a_{nj}A_{nj} = |\mathbf{A}|$$
$$a_{1j}A_{1k} + a_{2j}A_{2k} + \cdots + a_{nj}A_{nk} = 0 \quad \text{if } k \neq j$$

Expanding a determinant by a row or a column in terms of the cofactors of the same row or column, yields the determinant. Expanding by a row or a column in terms of the cofactors of a different row or column, yields 0.

20.7

- If all the elements in a row (or column) of $\mathbf{A}$ are 0, then $|\mathbf{A}| = 0$.
- If two rows (or two columns) of $\mathbf{A}$ are interchanged, the determinant changes sign but the absolute value remains unchanged.
- If all the elements in a single row (or column) of $\mathbf{A}$ are multiplied by a number c, the determinant is multiplied by c.
- If two of the rows (or columns) of $\mathbf{A}$ are proportional, then $|\mathbf{A}| = 0$.
- The value of $|\mathbf{A}|$ remains unchanged if a multiple of one row (or one column) is added to another row (or column).
- $|\mathbf{A}'| = |\mathbf{A}|$, where $\mathbf{A}'$ is the transpose of $\mathbf{A}$.

Important properties of determinants. $\mathbf{A}$ is a square matrix.

20.8

$$|\mathbf{AB}| = |\mathbf{A}| \cdot |\mathbf{B}|$$
$$|\mathbf{A} + \mathbf{B}| \neq |\mathbf{A}| + |\mathbf{B}| \quad \text{(in general)}$$

Properties of determinants. $\mathbf{A}$ and $\mathbf{B}$ are $n \times n$ matrices.

20.9

$$\begin{vmatrix} 1 & x_1 & x_1^2 \\ 1 & x_2 & x_2^2 \\ 1 & x_3 & x_3^2 \end{vmatrix} = (x_2 - x_1)(x_3 - x_1)(x_3 - x_2)$$

The Vandermonde determinant for $n = 3$.

20.10
$$\begin{vmatrix} 1 & x_1 & x_1^2 & \cdots & x_1^{n-1} \\ 1 & x_2 & x_2^2 & \cdots & x_2^{n-1} \\ 1 & x_3 & x_3^2 & \cdots & x_3^{n-1} \\ \vdots & \vdots & \vdots & \ddots & \vdots \\ 1 & x_n & x_n^2 & \cdots & x_n^{n-1} \end{vmatrix} = \prod_{1 \le j < i \le n} (x_i - x_j)$$

The general Vandermonde determinant.

20.11
$$\begin{vmatrix} a_1 & 1 & \cdots & 1 \\ 1 & a_2 & \cdots & 1 \\ \vdots & \vdots & \ddots & \vdots \\ 1 & 1 & \cdots & a_n \end{vmatrix}$$
$$= (a_1 - 1)(a_2 - 1) \cdots (a_n - 1) \left[1 + \sum_{i=1}^{n} \frac{1}{a_i - 1} \right]$$

A special determinant. $a_i \ne 1$ for $i = 1, \ldots, n$.

20.12
$$\begin{vmatrix} 0 & p_1 & \cdots & p_n \\ q_1 & a_{11} & \cdots & a_{1n} \\ \vdots & \vdots & \ddots & \vdots \\ q_n & a_{n1} & \cdots & a_{nn} \end{vmatrix} = - \sum_{i=1}^{n} \sum_{j=1}^{n} p_i A_{ji} q_j$$

A useful determinant ($n \ge 2$). A_{ji} is found in (20.5).

20.13
$$\begin{vmatrix} \alpha & p_1 & \cdots & p_n \\ q_1 & a_{11} & \cdots & a_{1n} \\ \vdots & \vdots & \ddots & \vdots \\ q_n & a_{n1} & \cdots & a_{nn} \end{vmatrix} = (\alpha - \mathbf{P}'\mathbf{A}^{-1}\mathbf{Q}) |\mathbf{A}|$$

Generalization of (20.12) when $\mathbf{A}^{-1}$ exists. $\mathbf{P}' = (p_1, \ldots, p_n)$, $\mathbf{Q}' = (q_1, \ldots, q_n)$.

20.14 $\quad |\mathbf{AB} + \mathbf{I}_m| = |\mathbf{BA} + \mathbf{I}_n|$

A useful result. $\mathbf{A}$ is $m \times n$, $\mathbf{B}$ is $n \times m$.

20.15
- A *minor* of order k in $\mathbf{A}$ is the determinant of a $k \times k$ matrix obtained by deleting all but k rows and all but k columns of $\mathbf{A}$.
- A *principal minor* of order k in $\mathbf{A}$ is a minor obtained by deleting all but k rows and all except the k columns with the *same* numbers.
- The *leading principal minor* of order k in $\mathbf{A}$ is the principal minor obtained by deleting all but the first k rows and the first k columns.

Definitions of minors, principal minors, and leading principal minors of a matrix.

20.16 $\quad D_k = \begin{vmatrix} a_{11} & a_{12} & \cdots & a_{1k} \\ a_{21} & a_{22} & \cdots & a_{2k} \\ \vdots & \vdots & \ddots & \vdots \\ a_{k1} & a_{k2} & \cdots & a_{kk} \end{vmatrix}, \quad k = 1, 2, \ldots, n$

The *leading principal minors* of $\mathbf{A} = (a_{ij})_{n \times n}$.

If $|\mathbf{A}| = |(a_{ij})_{n \times n}| \neq 0$, then the linear system of n equations and n unknowns,

$$a_{11}x_1 + a_{12}x_2 + \cdots + a_{1n}x_n = b_1$$
$$a_{21}x_1 + a_{22}x_2 + \cdots + a_{2n}x_n = b_2$$
$$\dots\dots\dots\dots\dots\dots\dots\dots\dots\dots\dots$$
$$a_{n1}x_1 + a_{n2}x_2 + \cdots + a_{nn}x_n = b_n$$

20.17 has the unique solution

$$x_j = \frac{|\mathbf{A}_j|}{|\mathbf{A}|}, \quad j = 1, 2, \ldots, n$$

where

$$|\mathbf{A}_j| = \begin{vmatrix} a_{11} & \cdots & a_{1j-1} & b_1 & a_{1j+1} & \cdots & a_{1n} \\ a_{21} & \cdots & a_{2j-1} & b_2 & a_{2j+1} & \cdots & a_{2n} \\ \vdots & & \vdots & \vdots & \vdots & & \vdots \\ a_{n1} & \cdots & a_{nj-1} & b_n & a_{nj+1} & \cdots & a_{nn} \end{vmatrix}$$

Cramer's rule. Note that $|\mathbf{A}_j|$ is obtained by replacing the jth column in $|\mathbf{A}|$ by the vector with components $b_1, b_2, \ldots, b_n$.

References

Most of the formulas are standard and can be found in almost any linear algebra text, e.g. Fraleigh and Beauregard (1995) or Lang (1987). See also Sydsæter and Hammond (2005). A standard reference is Gantmacher (1959).

Eigenvalues. Quadratic forms

21.1 A scalar λ is called an *eigenvalue* of an $n \times n$ matrix $\mathbf{A}$ if there exists an n-vector $\mathbf{c} \neq \mathbf{0}$ such that

$$\mathbf{Ac} = \lambda \mathbf{c}$$

The vector $\mathbf{c}$ is called an *eigenvector* of $\mathbf{A}$.

Eigenvalues and eigenvectors are also called *characteristic roots* and *characteristic vectors*. λ and $\mathbf{c}$ may be complex even if $\mathbf{A}$ is real.

21.2
$$|\mathbf{A} - \lambda\mathbf{I}| = \begin{vmatrix} a_{11} - \lambda & a_{12} & \cdots & a_{1n} \\ a_{21} & a_{22} - \lambda & \cdots & a_{2n} \\ \vdots & \vdots & \ddots & \vdots \\ a_{n1} & a_{n2} & \cdots & a_{nn} - \lambda \end{vmatrix}$$

The *eigenvalue polynomial* (the *characteristic polynomial*) of $\mathbf{A} = (a_{ij})_{n \times n}$. $\mathbf{I}$ is the unit matrix of order n.

21.3 λ is an eigenvalue of $\mathbf{A} \Leftrightarrow p(\lambda) = |\mathbf{A} - \lambda\mathbf{I}| = 0$

A necessary and sufficient condition for λ to be an eigenvalue of $\mathbf{A}$.

21.4
$$|\mathbf{A}| = \lambda_1 \cdot \lambda_2 \cdots \lambda_{n-1} \cdot \lambda_n$$
$$\text{tr}(\mathbf{A}) = a_{11} + \cdots + a_{nn} = \lambda_1 + \cdots + \lambda_n$$

$\lambda_1, \ldots, \lambda_n$ are the eigenvalues of $\mathbf{A}$.

21.5 Let $f(\ )$ be a polynomial. If λ is an eigenvalue of $\mathbf{A}$, then $f(\lambda)$ is an eigenvalue of $f(\mathbf{A})$.

Eigenvalues for matrix polynomials.

21.6 A square matrix $\mathbf{A}$ has an inverse if and only if 0 is not an eigenvalue of $\mathbf{A}$. If $\mathbf{A}$ has an inverse and λ is an eigenvalue of $\mathbf{A}$, then λ^{-1} is an eigenvalue of $\mathbf{A}^{-1}$.

How to find the eigenvalues of the inverse of a square matrix.

21.7 All eigenvalues of $\mathbf{A}$ have moduli (strictly) less than 1 if and only if $\mathbf{A}^t \to \mathbf{0}$ as $t \to \infty$.

An important result.

21.8 $\mathbf{AB}$ and $\mathbf{BA}$ have the same eigenvalues.

$\mathbf{A}$ and $\mathbf{B}$ are $n \times n$ matrices.

21.9 If $\mathbf{A}$ is symmetric and has only real elements, then all eigenvalues of $\mathbf{A}$ are reals.

21.10 If
$$p(\lambda) = |\mathbf{A} - \lambda\mathbf{I}| =$$
$$(-\lambda)^n + b_{n-1}(-\lambda)^{n-1} + \cdots + b_1(-\lambda) + b_0$$
is the eigenvalue polynomial of $\mathbf{A}$, then b_k is the sum of all principal minors of $\mathbf{A}$ of order $n - k$ (there are $\binom{n}{k}$ of them).

Characterization of the coefficients of the eigenvalue polynomial of an $n \times n$ matrix $\mathbf{A}$. (For principal minors, see (20.15).) $p(\lambda) = 0$ is called the *eigenvalue equation* or *characteristic equation* of $\mathbf{A}$.

21.11
$$\begin{vmatrix} a_{11} - \lambda & a_{12} \\ a_{21} & a_{22} - \lambda \end{vmatrix} = (-\lambda)^2 + b_1(-\lambda) + b_0$$
where $b_1 = a_{11} + a_{22} = \text{tr}(\mathbf{A})$, $b_0 = |\mathbf{A}|$

(21.10) for $n = 2$. ($\text{tr}(\mathbf{A})$ is the trace of $\mathbf{A}$.)

21.12
$$\begin{vmatrix} a_{11} - \lambda & a_{12} & a_{13} \\ a_{21} & a_{22} - \lambda & a_{23} \\ a_{31} & a_{32} & a_{33} - \lambda \end{vmatrix} =$$
$$(-\lambda)^3 + b_2(-\lambda)^2 + b_1(-\lambda) + b_0$$
where
$$b_2 = a_{11} + a_{22} + a_{33} = \text{tr}(\mathbf{A})$$
$$b_1 = \begin{vmatrix} a_{11} & a_{12} \\ a_{21} & a_{22} \end{vmatrix} + \begin{vmatrix} a_{11} & a_{13} \\ a_{31} & a_{33} \end{vmatrix} + \begin{vmatrix} a_{22} & a_{23} \\ a_{32} & a_{33} \end{vmatrix}$$
$$b_0 = |\mathbf{A}|$$

(21.10) for $n = 3$.

21.13 $\mathbf{A}$ is *diagonalizable* $\Leftrightarrow$ $\begin{cases} \mathbf{P}^{-1}\mathbf{A}\mathbf{P} = \mathbf{D} \text{ for} \\ \text{some matrix } \mathbf{P} \text{ and} \\ \text{some diagonal ma-} \\ \text{trix } \mathbf{D}. \end{cases}$

A definition.

21.14 $\mathbf{A}$ and $\mathbf{P}^{-1}\mathbf{A}\mathbf{P}$ have the same eigenvalues.

21.15 If $\mathbf{A} = (a_{ij})_{n \times n}$ has n distinct eigenvalues, then $\mathbf{A}$ is diagonalizable.

Sufficient (but NOT necessary) condition for $\mathbf{A}$ to be diagonalizable.

21.16 $\mathbf{A} = (a_{ij})_{n \times n}$ has n linearly independent eigenvectors, $\mathbf{x}_1, \ldots, \mathbf{x}_n$, with corresponding eigenvalues $\lambda_1, \ldots, \lambda_n$, if and only if
$$\mathbf{P}^{-1}\mathbf{A}\mathbf{P} = \begin{pmatrix} \lambda_1 & 0 & \cdots & 0 \\ 0 & \lambda_2 & \cdots & 0 \\ \vdots & \vdots & \ddots & \vdots \\ 0 & 0 & \cdots & \lambda_n \end{pmatrix}$$
where $\mathbf{P} = (\mathbf{x}_1, \ldots, \mathbf{x}_n)_{n \times n}$.

A characterization of diagonalizable matrices.

If $\mathbf{A} = (a_{ij})_{n \times n}$ is symmetric, with eigenvalues $\lambda_1, \lambda_2, \ldots, \lambda_n$, there exists an orthogonal matrix $\mathbf{U}$ such that

21.17
$$\mathbf{U}^{-1}\mathbf{A}\mathbf{U} = \begin{pmatrix} \lambda_1 & 0 & \cdots & 0 \\ 0 & \lambda_2 & \cdots & 0 \\ \vdots & \vdots & \ddots & \vdots \\ 0 & 0 & \cdots & \lambda_n \end{pmatrix}$$

The *spectral theorem* for symmetric matrices. For orthogonal matrices, see Chapter 22.

If $\mathbf{A}$ is an $n \times n$ matrix with eigenvalues $\lambda_1, \ldots, \lambda_n$ (not necessarily distinct), then there exists an invertible $n \times n$ matrix $\mathbf{T}$ such that

$$\mathbf{T}^{-1}\mathbf{A}\mathbf{T} = \begin{pmatrix} \mathbf{J}_{k_1}(\lambda_1) & 0 & \cdots & 0 \\ 0 & \mathbf{J}_{k_2}(\lambda_2) & \cdots & 0 \\ \vdots & \vdots & \ddots & \vdots \\ 0 & 0 & \cdots & \mathbf{J}_{k_r}(\lambda_r) \end{pmatrix}$$

21.18
where $k_1 + k_2 + \cdots + k_r = n$ and $\mathbf{J}_k$ is the $k \times k$ matrix

The *Jordan decomposition theorem.*

$$\mathbf{J}_k(\lambda) = \begin{pmatrix} \lambda & 1 & 0 & \cdots & 0 \\ 0 & \lambda & 1 & \cdots & 0 \\ \vdots & \vdots & \vdots & \ddots & \vdots \\ 0 & 0 & 0 & \cdots & 1 \\ 0 & 0 & 0 & \cdots & \lambda \end{pmatrix}, \quad \mathbf{J}_1(\lambda) = \lambda$$

21.19 Let $\mathbf{A}$ be a complex $n \times n$ matrix. Then there exists a unitary matrix $\mathbf{U}$ such that $\mathbf{U}^{-1}\mathbf{A}\mathbf{U}$ is upper triangular.

Schur's lemma. (For unitary matrices, see (19.50).)

21.20 Let $\mathbf{A} = (a_{ij})$ be a Hermitian matrix. Then there is a unitary matrix $\mathbf{U}$ such that $\mathbf{U}^{-1}\mathbf{A}\mathbf{U}$ is a diagonal matrix. All eigenvalues of $\mathbf{A}$ are then real.

The spectral theorem for Hermitian matrices. (For Hermitian matrices, see (19.50).)

21.21 Given any matrix $\mathbf{A} = (a_{ij})_{n \times n}$, there is for every $\varepsilon > 0$ a matrix $\mathbf{B}_\varepsilon = (b_{ij})_{n \times n}$, with n *distinct* eigenvalues, such that

$$\sum_{i,j=1}^{n} |a_{ij} - b_{ij}| < \varepsilon$$

By changing the elements of a matrix only slightly one gets a matrix with distinct eigenvalues.

21.22 A square matrix $\mathbf{A}$ satisfies its own eigenvalue equation:

$$p(\mathbf{A}) = (-\mathbf{A})^n + b_{n-1}(-\mathbf{A})^{n-1}$$
$$+ \cdots + b_1(-\mathbf{A}) + b_0\mathbf{I} = \mathbf{0}$$

The *Cayley–Hamilton* theorem. The polynomial $p(\ )$ is defined in (21.10).

21.23 $\quad \mathbf{A} = \begin{pmatrix} a_{11} & a_{12} \\ a_{21} & a_{22} \end{pmatrix} \Rightarrow \mathbf{A}^2 - \text{tr}\,(\mathbf{A})\mathbf{A} + |\mathbf{A}|\mathbf{I} = \mathbf{0}$

The Cayley–Hamilton theorem for $n = 2$. (See (21.11).)

21.24

$$Q = \sum_{i=1}^{n} \sum_{j=1}^{n} a_{ij} x_i x_j =$$

$$a_{11} x_1^2 + a_{12} x_1 x_2 + \cdots + a_{1n} x_1 x_n$$
$$+ a_{21} x_2 x_1 + a_{22} x_2^2 + \cdots + a_{2n} x_2 x_n$$
$$+ \cdots\cdots\cdots\cdots\cdots\cdots\cdots\cdots\cdots\cdots$$
$$+ a_{n1} x_n x_1 + a_{n2} x_n x_2 + \cdots + a_{nn} x_n^2$$

A *quadratic form* in n variables $x_1, \ldots, x_n$. One can assume, without loss of generality, that $a_{ij} = a_{ji}$ for all $i, j = 1, \ldots, n$.

21.25

$$Q = \sum_{i=1}^{n} \sum_{j=1}^{n} a_{ij} x_i x_j = \mathbf{x}'\mathbf{A}\mathbf{x}, \quad \text{where}$$

$$\mathbf{x} = \begin{pmatrix} x_1 \\ \vdots \\ x_n \end{pmatrix} \text{ and } \mathbf{A} = \begin{pmatrix} a_{11} & a_{12} & \cdots & a_{1n} \\ a_{21} & a_{22} & \cdots & a_{2n} \\ \vdots & \vdots & \ddots & \vdots \\ a_{n1} & a_{n2} & \cdots & a_{nn} \end{pmatrix}$$

A quadratic form in matrix formulation. One can assume, without loss of generality, that $\mathbf{A}$ is symmetric.

21.26

$\mathbf{x}'\mathbf{A}\mathbf{x}$ is PD $\Leftrightarrow$ $\mathbf{x}'\mathbf{A}\mathbf{x} > 0$ for all $\mathbf{x} \neq \mathbf{0}$

$\mathbf{x}'\mathbf{A}\mathbf{x}$ is PSD $\Leftrightarrow$ $\mathbf{x}'\mathbf{A}\mathbf{x} \geq 0$ for all $\mathbf{x}$

$\mathbf{x}'\mathbf{A}\mathbf{x}$ is ND $\Leftrightarrow$ $\mathbf{x}'\mathbf{A}\mathbf{x} < 0$ for all $\mathbf{x} \neq \mathbf{0}$

$\mathbf{x}'\mathbf{A}\mathbf{x}$ is NSD $\Leftrightarrow$ $\mathbf{x}'\mathbf{A}\mathbf{x} \leq 0$ for all $\mathbf{x}$

$\mathbf{x}'\mathbf{A}\mathbf{x}$ is ID $\Leftrightarrow$ $\mathbf{x}'\mathbf{A}\mathbf{x}$ is neither PSD nor NSD

Definiteness types for quadratic forms ($\mathbf{x}'\mathbf{A}\mathbf{x}$) and symmetric matrices ($\mathbf{A}$). The five types are: *positive definite* (PD), *positive semidefinite* (PSD), *negative definite* (ND), *negative semidefinite* (NSD), and *indefinite* (ID).

21.27

$\mathbf{x}'\mathbf{A}\mathbf{x}$ is PD $\Rightarrow$ $a_{ii} > 0$ for $i = 1, \ldots, n$

$\mathbf{x}'\mathbf{A}\mathbf{x}$ is PSD $\Rightarrow$ $a_{ii} \geq 0$ for $i = 1, \ldots, n$

$\mathbf{x}'\mathbf{A}\mathbf{x}$ is ND $\Rightarrow$ $a_{ii} < 0$ for $i = 1, \ldots, n$

$\mathbf{x}'\mathbf{A}\mathbf{x}$ is NSD $\Rightarrow$ $a_{ii} \leq 0$ for $i = 1, \ldots, n$

Let $x_i = 1$ and $x_j = 0$ for $j \neq i$ in (21.24).

21.28

$\mathbf{x}'\mathbf{A}\mathbf{x}$ is PD $\Leftrightarrow$ all eigenvalues of $\mathbf{A}$ are > 0

$\mathbf{x}'\mathbf{A}\mathbf{x}$ is PSD $\Leftrightarrow$ all eigenvalues of $\mathbf{A}$ are ≥ 0

$\mathbf{x}'\mathbf{A}\mathbf{x}$ is ND $\Leftrightarrow$ all eigenvalues of $\mathbf{A}$ are < 0

$\mathbf{x}'\mathbf{A}\mathbf{x}$ is NSD $\Leftrightarrow$ all eigenvalues of $\mathbf{A}$ are ≤ 0

A characterization of definite quadratic forms (matrices) in terms of the signs of the eigenvalues.

21.29

$\mathbf{x}'\mathbf{A}\mathbf{x}$ is indefinite (ID) if and only if $\mathbf{A}$ has at least one positive and one negative eigenvalue.

A characterization of *indefinite* quadratic forms.

21.30

$\mathbf{x}'\mathbf{Ax}$ is PD $\Leftrightarrow D_k > 0$ for $k = 1, \ldots, n$

$\mathbf{x}'\mathbf{Ax}$ is ND $\Leftrightarrow (-1)^k D_k > 0$ for $k = 1, \ldots, n$

where the *leading principal minors* D_k of $\mathbf{A}$ are

$$D_k = \begin{vmatrix} a_{11} & a_{12} & \cdots & a_{1k} \\ a_{21} & a_{22} & \cdots & a_{2k} \\ \vdots & \vdots & \ddots & \vdots \\ a_{k1} & a_{k2} & \cdots & a_{kk} \end{vmatrix}, \quad k = 1, 2, \ldots, n$$

A characterization of definite quadratic forms (matrices) in terms of leading principal minors. *Note that replacing $>$ by $\geq$ will NOT give criteria for the semidefinite case.* Example: $Q = 0x_1^2 + 0x_1x_2 - x_2^2$.

21.31

$\mathbf{x}'\mathbf{Ax}$ is PSD $\Leftrightarrow \Delta_r \geq 0$ for $r = 1, \ldots, n$

$\mathbf{x}'\mathbf{Ax}$ is NSD $\Leftrightarrow (-1)^r \Delta_r \geq 0$ for $r = 1, \ldots, n$

For each r, Δ_r runs through all principal minors of $\mathbf{A}$ of order r.

Characterizations of positive and negative semidefinite quadratic forms (matrices) in terms of principal minors. (For principal minors, see (20.15).)

21.32

If $\mathbf{A} = (a_{ij})_{n \times n}$ is positive definite and $\mathbf{P}$ is $n \times m$ with $r(\mathbf{P}) = m$, then $\mathbf{P}'\mathbf{AP}$ is positive definite.

Results on positive definite matrices.

21.33

If $\mathbf{P}$ is $n \times m$ and $r(\mathbf{P}) = m$, then $\mathbf{P}'\mathbf{P}$ is positive definite and has rank m.

21.34

If $\mathbf{A}$ is positive definite, there exists a non-singular matrix $\mathbf{P}$ such that $\mathbf{PAP}' = \mathbf{I}$ and $\mathbf{P}'\mathbf{P} = \mathbf{A}^{-1}$.

21.35

Let $\mathbf{A}$ be an $m \times n$ matrix with $r(\mathbf{A}) = k$. Then there exist a unitary $m \times m$ matrix $\mathbf{U}$, a unitary $n \times n$ matrix $\mathbf{V}$, and a $k \times k$ diagonal matrix $\mathbf{D}$, with only strictly positive diagonal elements, such that

$$\mathbf{A} = \mathbf{USV}^*, \text{ where } \mathbf{S} = \begin{pmatrix} \mathbf{D} & \mathbf{0} \\ \mathbf{0} & \mathbf{0} \end{pmatrix}$$

If $k = m = n$, then $\mathbf{S} = \mathbf{D}$. If $\mathbf{A}$ is real, $\mathbf{U}$ and $\mathbf{V}$ can be chosen as real, orthogonal matrices.

The *singular value decomposition theorem*. The diagonal elements of $\mathbf{D}$ are called *singular values* for the matrix $\mathbf{A}$. Unitary matrices are defined in (19.50), and orthogonal matrices are defined in (22.8).

21.36

Let $\mathbf{A}$ and $\mathbf{B}$ be symmetric $n \times n$ matrices. Then there exists an orthogonal matrix $\mathbf{Q}$ such that both $\mathbf{Q}'\mathbf{AQ}$ and $\mathbf{Q}'\mathbf{BQ}$ are diagonal matrices, if and only if $\mathbf{AB} = \mathbf{BA}$.

Simultaneous diagonalization.

The quadratic form

$$(*) \quad Q = \sum_{i=1}^{n} \sum_{j=1}^{n} a_{ij} x_i x_j, \qquad (a_{ij} = a_{ji})$$

21.37

is *positive (negative) definite subject to the linear constraints*

$$b_{11}x_1 + \cdots + b_{1n}x_n = 0$$

$$(**) \quad \dots\dots\dots\dots\dots\dots \qquad (m < n)$$

$$b_{m1}x_1 + \cdots + b_{mn}x_n = 0$$

if $Q > 0$ (< 0) for all $(x_1, \dots, x_n) \neq (0, \dots, 0)$ that satisfy $(**)$.

A definition of positive (negative) definiteness subject to linear constraints.

21.38

$$D_r = \begin{vmatrix} 0 & \cdots & 0 & b_{11} & \cdots & b_{1r} \\ \vdots & \ddots & \vdots & \vdots & & \vdots \\ 0 & \cdots & 0 & b_{m1} & \cdots & b_{mr} \\ b_{11} & \cdots & b_{m1} & a_{11} & \cdots & a_{1r} \\ \vdots & & \vdots & \vdots & \ddots & \vdots \\ b_{1r} & \cdots & b_{mr} & a_{r1} & \cdots & a_{rr} \end{vmatrix}$$

A bordered determinant associated with (21.37), $r = 1, \dots, n$.

21.39

Necessary and sufficient conditions for the quadratic form $(*)$ in (21.37) to be positive definite subject to the constraints $(**)$, assuming that the *first m columns* of the matrix $(b_{ij})_{m \times n}$ are linearly independent, is that

$$(-1)^m D_r > 0, \quad r = m+1, \dots, n$$

The corresponding conditions for $(*)$ to be negative definite subject to the constraints $(**)$ is that

$$(-1)^r D_r > 0, \quad r = m+1, \dots, n$$

A test for definiteness of quadratic forms subject to linear constraints. (Assuming that the rank of $(b_{ij})_{m \times n}$ is m is not enough, as is shown by the example, $Q(x_1, x_2, x_3) = x_1^2 + x_2^2 - x_3^2$ with the constraint $x_3 = 0$.)

21.40

The quadratic form $ax^2 + 2bxy + cy^2$ is positive for all $(x, y) \neq (0, 0)$ satisfying the constraint $px + qy = 0$, if and only if

$$\begin{vmatrix} 0 & p & q \\ p & a & b \\ q & b & c \end{vmatrix} < 0$$

A special case of (21.39), assuming $(p, q) \neq (0, 0)$.

References

Most of the formulas can be found in almost any linear algebra text, e.g. Fraleigh and Beauregard (1995) or Lang (1987). See also Horn and Johnson (1985) and Sydsæter et al. (2005). Gantmacher (1959) is a standard reference.

Chapter 22

Special matrices. Leontief systems

Idempotent matrices

22.1 $\mathbf{A} = (a_{ij})_{n \times n}$ is *idempotent* $\iff \mathbf{A}^2 = \mathbf{A}$

Definition of an idempotent matrix.

22.2 $\mathbf{A}$ is idempotent $\iff \mathbf{I} - \mathbf{A}$ is idempotent.

Properties of idempotent matrices.

22.3 $\mathbf{A}$ is idempotent $\Rightarrow$ 0 and 1 are the only possible eigenvalues, and $\mathbf{A}$ is positive semidefinite.

22.4 $\mathbf{A}$ is idempotent with k eigenvalues equal to 1 $\Rightarrow r(\mathbf{A}) = \mathrm{tr}(\mathbf{A}) = k$.

22.5 $\mathbf{A}$ is idempotent and $\mathbf{C}$ is orthogonal $\Rightarrow \mathbf{C}'\mathbf{A}\mathbf{C}$ is idempotent.

An orthogonal matrix is defined in (22.8).

22.6 $\mathbf{A}$ is idempotent $\iff$ its associated linear transformation is a projection.

A linear transformation P from $\mathbb{R}^n$ into $\mathbb{R}^n$ is a *projection* if $P(P(\mathbf{x})) = P(\mathbf{x})$ for all $\mathbf{x}$ in $\mathbb{R}^n$.

22.7 $\mathbf{I}_n - \mathbf{X}(\mathbf{X}'\mathbf{X})^{-1}\mathbf{X}'$ is idempotent.

$\mathbf{X}$ is $n \times m$, $|\mathbf{X}'\mathbf{X}| \neq 0$.

Orthogonal matrices

22.8 $\mathbf{P} = (p_{ij})_{n \times n}$ is *orthogonal* $\iff \mathbf{P}'\mathbf{P} = \mathbf{P}\mathbf{P}' = \mathbf{I}_n$

Definition of an orthogonal matrix.

22.9 $\mathbf{P}$ is orthogonal $\iff$ the column vectors of $\mathbf{P}$ are mutually orthogonal unit vectors.

A property of orthogonal matrices.

22.10   **P** and **Q** are orthogonal $\Rightarrow$ **PQ** is orthogonal. | Properties of orthogonal matrices.

22.11   **P** orthogonal $\Rightarrow$ $|\mathbf{P}| = \pm 1$, and 1 and -1 are the only possible real eigenvalues.

22.12   **P** orthogonal $\Leftrightarrow$ $\|\mathbf{Px}\| = \|\mathbf{x}\|$ for all **x** in $\mathbb{R}^n$. | Orthogonal transformations preserve lengths of vectors.

22.13   If **P** is orthogonal, the angle between **Px** and **Py** equals the angle between **x** and **y**. | Orthogonal transformations preserve angles.

Permutation matrices

22.14   **P** $= (p_{ij})_{n \times n}$ is a *permutation* matrix if in each row and each column of **P** there is one element equal to 1 and the rest of the elements are 0. | Definition of a permutation matrix.

22.15   **P** is a permutation matrix $\Rightarrow$ **P** is nonsingular and orthogonal. | Properties of permutation matrices.

Nonnegative matrices

22.16 $\mathbf{A} = (a_{ij})_{m \times n} \geq \mathbf{0} \iff a_{ij} \geq 0$ for all i, j
$\mathbf{A} = (a_{ij})_{m \times n} > \mathbf{0} \iff a_{ij} > 0$ for all i, j | Definitions of *nonnegative* and *positive* matrices.

22.17   If $\mathbf{A} = (a_{ij})_{n \times n} \geq 0$, **A** has at least one nonnegative eigenvalue. The largest nonnegative eigenvalue is called the *Frobenius root* of **A** and it is denoted by $\lambda(\mathbf{A})$. **A** has a nonnegative eigenvector corresponding to $\lambda(\mathbf{A})$. | Definition of the Frobenius root (or *dominant root*) of a nonnegative matrix.

22.18
- μ is an eigenvalue of **A** $\Rightarrow |\mu| \leq \lambda(\mathbf{A})$
- $0 \leq \mathbf{A}_1 \leq \mathbf{A}_2 \Rightarrow \lambda(\mathbf{A}_1) \leq \lambda(\mathbf{A}_2)$
- $\rho > \lambda(\mathbf{A}) \iff (\rho\mathbf{I} - \mathbf{A})^{-1}$ exists and is ≥ 0
- $\min\limits_{1 \leq j \leq n} \sum\limits_{i=1}^{n} a_{ij} \leq \lambda(\mathbf{A}) \leq \max\limits_{1 \leq j \leq n} \sum\limits_{i=1}^{n} a_{ij}$

| Properties of nonnegative matrices. $\lambda(\mathbf{A})$ is the Frobenius root of *A*.

22.19 The matrix $\mathbf{A} = (a_{ij})_{n \times n}$ is *decomposable* or *reducible* if by interchanging some rows and the corresponding columns it is possible to transform the matrix $\mathbf{A}$ to

$$\begin{pmatrix} \mathbf{A}_{11} & \mathbf{A}_{12} \\ \mathbf{0} & \mathbf{A}_{22} \end{pmatrix}$$

where $\mathbf{A}_{11}$ and $\mathbf{A}_{22}$ are square submatrices.

Definition of a decomposable square matrix. A matrix that is not decomposable (reducible) is called *indecomposable* (*irreducible*).

22.20 $\mathbf{A} = (a_{ij})_{n \times n}$ is decomposable if and only if there exists a permutation matrix $\mathbf{P}$ such that

$$\mathbf{P}^{-1}\mathbf{A}\mathbf{P} = \begin{pmatrix} \mathbf{A}_{11} & \mathbf{A}_{12} \\ \mathbf{0} & \mathbf{A}_{22} \end{pmatrix}$$

where $\mathbf{A}_{11}$ and $\mathbf{A}_{22}$ are square submatrices.

A characterization of decomposable matrices.

22.21 If $\mathbf{A} = (a_{ij})_{n \times n} \geq \mathbf{0}$ is indecomposable, then
- the Frobenius root $\lambda(\mathbf{A})$ is > 0, it is a simple root of the eigenvalue equation, and there exists an associated eigenvector $\mathbf{x} > \mathbf{0}$.
- If $\mathbf{A}\mathbf{x} = \mu\mathbf{x}$ for some $\mu \geq 0$ and $\mathbf{x} > \mathbf{0}$, then $\mu = \lambda(\mathbf{A})$.

Properties of indecomposable matrices.

22.22 $\mathbf{A} = (a_{ij})_{n \times n}$ has a *dominant diagonal* (d.d.) if there exist positive numbers $d_1, \ldots, d_n$ such that

$$d_j|a_{jj}| > \sum_{i \neq j} d_i|a_{ij}| \quad \text{for} \quad j = 1, \ldots, n$$

Definition of a dominant diagonal matrix.

22.23 Suppose $\mathbf{A}$ has a dominant diagonal. Then:
- $|\mathbf{A}| \neq 0$.
- If the diagonal elements are all positive, then all the eigenvalues of $\mathbf{A}$ have positive real parts.

Properties of dominant diagonal matrices.

Leontief systems

22.24 If $\mathbf{A} = (a_{ij})_{n \times n} \geq \mathbf{0}$ and $\mathbf{c} \geq \mathbf{0}$, then

$$\mathbf{A}\mathbf{x} + \mathbf{c} = \mathbf{x}$$

is called a *Leontief system*.

Definition of a Leontief system. $\mathbf{x}$ and $\mathbf{c}$ are $n \times 1$-matrices.

22.25 If $\sum_{i=1}^{n} a_{ij} < 1$ for $j = 1, \ldots, n$, then the Leontief system has a solution $\mathbf{x} \geq \mathbf{0}$.

Sufficient condition for a Leontief system to have a nonnegative solution.

The Leontief system $\mathbf{A}\mathbf{x} + \mathbf{c} = \mathbf{x}$ has a solution $\mathbf{x} \geq \mathbf{0}$ for every $\mathbf{c} \geq \mathbf{0}$, if and only if one (and hence all) of the following equivalent conditions is satisfied:

- The matrix $(\mathbf{I} - \mathbf{A})^{-1}$ exists, is nonnegative, and is equal to $\mathbf{I} + \mathbf{A} + \mathbf{A}^2 + \cdots$.

22.26

- $\mathbf{A}^m \to \mathbf{0}$ as $m \to \infty$.

- Every eigenvalue of $\mathbf{A}$ has modulus < 1.

- $$\begin{vmatrix} 1 - a_{11} & -a_{12} & \cdots & -a_{1k} \\ -a_{21} & 1 - a_{22} & \cdots & -a_{2k} \\ \vdots & \vdots & \ddots & \vdots \\ -a_{k1} & -a_{k2} & \cdots & 1 - a_{kk} \end{vmatrix} > 0$$

  for $k = 1, \ldots, n$.

Necessary and sufficient conditions for the Leontief system to have a nonnegative solution. The last conditions are the *Hawkins–Simon conditions*.

22.27

If $0 \leq a_{ii} < 1$ for $i = 1, \ldots, n$, and $a_{ij} \geq 0$ for all $i \neq j$, then the system $\mathbf{A}\mathbf{x} + \mathbf{c} = \mathbf{x}$ will have a solution $\mathbf{x} \geq \mathbf{0}$ for every $\mathbf{c} \geq \mathbf{0}$ if and only if $\mathbf{I} - \mathbf{A}$ has a dominant diagonal.

A necessary and sufficient condition for the Leontief system to have a nonnegative solution.

References

For the matrix results see Gantmacher (1959) or Horn and Johnson (1985). For Leontief systems, see Nikaido (1970) and Takayama (1985).

Chapter 23

Kronecker products and the vec operator. Differentiation of vectors and matrices

23.1 $\quad \mathbf{A} \otimes \mathbf{B} = \begin{pmatrix} a_{11}\mathbf{B} & a_{12}\mathbf{B} & \cdots & a_{1n}\mathbf{B} \\ a_{21}\mathbf{B} & a_{22}\mathbf{B} & \cdots & a_{2n}\mathbf{B} \\ \vdots & \vdots & & \vdots \\ a_{m1}\mathbf{B} & a_{m2}\mathbf{B} & \cdots & a_{mn}\mathbf{B} \end{pmatrix}$

The *Kronecker product* of $\mathbf{A} = (a_{ij})_{m \times n}$ and $\mathbf{B} = (b_{ij})_{p \times q}$. $\mathbf{A} \otimes \mathbf{B}$ is $mp \times nq$. In general, the Kronecker product is not commutative, $\mathbf{A} \otimes \mathbf{B} \neq \mathbf{B} \otimes \mathbf{A}$.

23.2 $\quad \begin{pmatrix} a_{11} & a_{12} \\ a_{21} & a_{22} \end{pmatrix} \otimes \begin{pmatrix} b_{11} & b_{12} \\ b_{21} & b_{22} \end{pmatrix} =$

$\begin{pmatrix} a_{11}b_{11} & a_{11}b_{12} & a_{12}b_{11} & a_{12}b_{12} \\ a_{11}b_{21} & a_{11}b_{22} & a_{12}b_{21} & a_{12}b_{22} \\ a_{21}b_{11} & a_{21}b_{12} & a_{22}b_{11} & a_{22}b_{12} \\ a_{21}b_{21} & a_{21}b_{22} & a_{22}b_{21} & a_{22}b_{22} \end{pmatrix}$

A special case of (23.1).

23.3 $\quad \mathbf{A} \otimes \mathbf{B} \otimes \mathbf{C} = (\mathbf{A} \otimes \mathbf{B}) \otimes \mathbf{C} = \mathbf{A} \otimes (\mathbf{B} \otimes \mathbf{C})$

Valid in general.

23.4 $\quad (\mathbf{A} + \mathbf{B}) \otimes (\mathbf{C} + \mathbf{D}) =$

$\quad\quad\quad \mathbf{A} \otimes \mathbf{C} + \mathbf{A} \otimes \mathbf{D} + \mathbf{B} \otimes \mathbf{C} + \mathbf{B} \otimes \mathbf{D}$

Valid if $\mathbf{A} + \mathbf{B}$ and $\mathbf{C} + \mathbf{D}$ are defined.

23.5 $\quad (\mathbf{A} \otimes \mathbf{B})(\mathbf{C} \otimes \mathbf{D}) = \mathbf{AC} \otimes \mathbf{BD}$

Valid if $\mathbf{AC}$ and $\mathbf{BD}$ are defined.

23.6 $\quad (\mathbf{A} \otimes \mathbf{B})' = \mathbf{A}' \otimes \mathbf{B}'$

Rule for transposing a Kronecker product.

23.7 $\quad (\mathbf{A} \otimes \mathbf{B})^{-1} = \mathbf{A}^{-1} \otimes \mathbf{B}^{-1}$

Valid if $\mathbf{A}^{-1}$ and $\mathbf{B}^{-1}$ exist.

23.8 $\quad \operatorname{tr}(\mathbf{A} \otimes \mathbf{B}) = \operatorname{tr}(\mathbf{A})\operatorname{tr}(\mathbf{B})$

$\mathbf{A}$ and $\mathbf{B}$ are square matrices, not necessarily of the same order.

23.9 $\quad \alpha \otimes \mathbf{A} = \alpha \mathbf{A} = \mathbf{A}\alpha = \mathbf{A} \otimes \alpha$ | α is a 1×1 scalar matrix.

23.10 If $\lambda_1, \ldots, \lambda_n$ are the eigenvalues of $\mathbf{A}$, and if $\mu_1, \ldots, \mu_p$ are the eigenvalues of $\mathbf{B}$, then the np eigenvalues of $\mathbf{A} \otimes \mathbf{B}$ are $\lambda_i \mu_j$, $i = 1, \ldots, n$, $j = 1, \ldots, p$. | The eigenvalues of $\mathbf{A} \otimes \mathbf{B}$, where $\mathbf{A}$ is $n \times n$ and $\mathbf{B}$ is $p \times p$.

23.11 If $\mathbf{x}$ is an eigenvector of $\mathbf{A}$, and $\mathbf{y}$ is an eigenvector for $\mathbf{B}$, then $\mathbf{x} \otimes \mathbf{y}$ is an eigenvector of $\mathbf{A} \otimes \mathbf{B}$. | NOTE: An eigenvector of $\mathbf{A} \otimes \mathbf{B}$ is not necessarily the Kronecker product of an eigenvector of $\mathbf{A}$ and an eigenvector of $\mathbf{B}$.

23.12 If $\mathbf{A}$ and $\mathbf{B}$ are positive (semi-)definite, then $\mathbf{A} \otimes \mathbf{B}$ is positive (semi-)definite. | Follows from (23.10).

23.13 $\quad |\mathbf{A} \otimes \mathbf{B}| = |\mathbf{A}|^p \cdot |\mathbf{B}|^n$ | $\mathbf{A}$ is $n \times n$, $\mathbf{B}$ is $p \times p$.

23.14 $\quad r(\mathbf{A} \otimes \mathbf{B}) = r(\mathbf{A})\, r(\mathbf{B})$ | The rank of a Kronecker product.

23.15 If $\mathbf{A} = (\mathbf{a}_1, \mathbf{a}_2, \ldots, \mathbf{a}_n)_{m \times n}$, then

$$\mathrm{vec}(\mathbf{A}) = \begin{pmatrix} \mathbf{a}_1 \\ \mathbf{a}_2 \\ \vdots \\ \mathbf{a}_n \end{pmatrix}_{mn \times 1}$$

| $\mathrm{vec}(\mathbf{A})$ consists of the columns of $\mathbf{A}$ placed below each other.

23.16 $\quad \mathrm{vec} \begin{pmatrix} a_{11} & a_{12} \\ a_{21} & a_{22} \end{pmatrix} = \begin{pmatrix} a_{11} \\ a_{21} \\ a_{12} \\ a_{22} \end{pmatrix}$ | A special case of (23.15).

23.17 $\quad \mathrm{vec}(\mathbf{A} + \mathbf{B}) = \mathrm{vec}(\mathbf{A}) + \mathrm{vec}(\mathbf{B})$ | Valid if $\mathbf{A} + \mathbf{B}$ is defined.

23.18 $\quad \mathrm{vec}(\mathbf{ABC}) = (\mathbf{C}' \otimes \mathbf{A})\, \mathrm{vec}(\mathbf{B})$ | Valid if the product $\mathbf{ABC}$ is defined.

23.19 $\quad \mathrm{tr}(\mathbf{AB}) = (\mathrm{vec}(\mathbf{A}'))'\, \mathrm{vec}(\mathbf{B}) = (\mathrm{vec}(\mathbf{B}'))'\, \mathrm{vec}(\mathbf{A})$ | Valid if the operations are defined.

Differentiation of vectors and matrices

23.20
$$\frac{\partial y}{\partial \mathbf{x}} = \left(\frac{\partial y}{\partial x_1}, \ldots, \frac{\partial y}{\partial x_n} \right)$$

If $y = f(x_1, \ldots, x_n) = f(\mathbf{x})$, then

The gradient of $y = f(\mathbf{x})$. (The derivative of a scalar function w.r.t. a vector variable.) An alternative notation for the gradient is $\nabla f(\mathbf{x})$. See (4.26).

23.21
$$y_1 = f_1(x_1, \ldots, x_n)$$
$$\ldots\ldots\ldots\ldots\ldots \iff \mathbf{y} = \mathbf{f}(\mathbf{x})$$
$$y_m = f_m(x_1, \ldots, x_n)$$

A transformation $\mathbf{f}$ from $\mathbb{R}^n$ to $\mathbb{R}^m$. We let $\mathbf{x}$ and $\mathbf{y}$ be column vectors.

23.22
$$\frac{\partial \mathbf{y}}{\partial \mathbf{x}} = \begin{pmatrix} \dfrac{\partial y_1(\mathbf{x})}{\partial x_1} & \cdots & \dfrac{\partial y_1(\mathbf{x})}{\partial x_n} \\ \vdots & & \vdots \\ \dfrac{\partial y_m(\mathbf{x})}{\partial x_1} & \cdots & \dfrac{\partial y_m(\mathbf{x})}{\partial x_n} \end{pmatrix}$$

The *Jacobian matrix* of the transformation in (23.21). (The derivative of a vector function w.r.t. a vector variable.)

23.23
$$\frac{\partial^2 \mathbf{y}}{\partial \mathbf{x} \partial \mathbf{x}'} = \frac{\partial}{\partial \mathbf{x}} \operatorname{vec} \left[\left(\frac{\partial \mathbf{y}}{\partial \mathbf{x}} \right)' \right]$$

For the vec operator, see (23.15).

23.24
$$\frac{\partial \mathbf{A}(\mathbf{r})}{\partial \mathbf{r}} = \frac{\partial}{\partial \mathbf{r}} \operatorname{vec}(\mathbf{A}(\mathbf{r}))$$

A general definition of the derivative of a matrix w.r.t. a vector.

23.25
$$\frac{\partial^2 y}{\partial \mathbf{x} \partial \mathbf{x}'} = \begin{pmatrix} \dfrac{\partial^2 y}{\partial x_1^2} & \cdots & \dfrac{\partial^2 y}{\partial x_n \partial x_1} \\ \vdots & \ddots & \vdots \\ \dfrac{\partial^2 y}{\partial x_1 \partial x_n} & \cdots & \dfrac{\partial^2 y}{\partial x_n^2} \end{pmatrix}$$

A special case of (23.23). ($\partial^2 y/\partial \mathbf{x} \partial \mathbf{x}'$ is the Hessian matrix defined in (13.24).)

23.26
$$\frac{\partial}{\partial \mathbf{x}}(\mathbf{a}' \cdot \mathbf{x}) = \mathbf{a}'$$

$\mathbf{a}$ and $\mathbf{x}$ are $n \times 1$-vectors.

23.27
$$\frac{\partial}{\partial \mathbf{x}}(\mathbf{x}' \mathbf{A} \mathbf{x}) = \mathbf{x}'(\mathbf{A} + \mathbf{A}')$$
$$\frac{\partial^2}{\partial \mathbf{x} \partial \mathbf{x}'}(\mathbf{x}' \mathbf{A} \mathbf{x}) = \mathbf{A} + \mathbf{A}'$$

Differentiation of a quadratic form. $\mathbf{A}$ is $n \times n$, $\mathbf{x}$ is $n \times 1$.

23.28
$$\frac{\partial}{\partial \mathbf{x}}(\mathbf{A} \mathbf{x}) = \mathbf{A}$$

$\mathbf{A}$ is $m \times n$, $\mathbf{x}$ is $n \times 1$.

If $\mathbf{y} = \mathbf{A}(\mathbf{r})\mathbf{x}(\mathbf{r})$, then

23.29
$$\frac{\partial \mathbf{y}}{\partial \mathbf{r}} = (\mathbf{x}' \otimes \mathbf{I}_m)\frac{\partial \mathbf{A}}{\partial \mathbf{r}} + \mathbf{A}\frac{\partial \mathbf{x}}{\partial \mathbf{r}}$$

$\mathbf{A}(\mathbf{r})$ is $m \times n$, $\mathbf{x}(\mathbf{r})$ is $n \times 1$ and $\mathbf{r}$ is $k \times 1$.

If $y = f(\mathbf{A})$, then

23.30
$$\frac{\partial y}{\partial \mathbf{A}} = \begin{pmatrix} \dfrac{\partial y}{\partial a_{11}} & \cdots & \dfrac{\partial y}{\partial a_{1n}} \\ \vdots & & \vdots \\ \dfrac{\partial y}{\partial a_{m1}} & \cdots & \dfrac{\partial y}{\partial a_{mn}} \end{pmatrix}$$

Definition of the derivative of a scalar function of an $m \times n$ matrix $\mathbf{A} = (a_{ij})$.

23.31
$$\frac{\partial |\mathbf{A}|}{\partial \mathbf{A}} = (A_{ij}) = |\mathbf{A}|(\mathbf{A}')^{-1}$$

$\mathbf{A}$ is $n \times n$. (A_{ij}) is the matrix of cofactors of $\mathbf{A}$. (See (19.16).) The last equality holds if $\mathbf{A}$ is invertible.

23.32
$$\frac{\partial \operatorname{tr}(\mathbf{A})}{\partial \mathbf{A}} = \mathbf{I}_n, \qquad \frac{\partial \operatorname{tr}(\mathbf{A}'\mathbf{A})}{\partial \mathbf{A}} = 2\mathbf{A}$$

$\mathbf{A}$ is $n \times n$. $\operatorname{tr}(\mathbf{A})$ is the trace of $\mathbf{A}$.

23.33
$$\frac{\partial a^{ij}}{\partial a_{hk}} = -a^{ih}a^{kj}; \qquad i, j, h, k = 1, \ldots, n$$

a^{ij} is the (i, j)th element of $\mathbf{A}^{-1}$.

References

The definitions above are common in the economic literature, see Dhrymes (1978). Magnus and Neudecker (1988) and Lütkepohl (1996) develop a more consistent notation and have all the results quoted here and many more.

Chapter 24

Comparative statics

24.1
$$E_1(\mathbf{p}, \mathbf{a}) = S_1(\mathbf{p}, \mathbf{a}) - D_1(\mathbf{p}, \mathbf{a})$$
$$E_2(\mathbf{p}, \mathbf{a}) = S_2(\mathbf{p}, \mathbf{a}) - D_2(\mathbf{p}, \mathbf{a})$$
$$\dots\dots\dots\dots\dots\dots\dots\dots\dots\dots\dots\dots$$
$$E_n(\mathbf{p}, \mathbf{a}) = S_n(\mathbf{p}, \mathbf{a}) - D_n(\mathbf{p}, \mathbf{a})$$

$S_i(\mathbf{p}, \mathbf{a})$ is supply and $D_i(\mathbf{p}, \mathbf{a})$ is demand for good i. $E_i(\mathbf{p}, \mathbf{a})$ is excess supply. $\mathbf{p} = (p_1, \dots, p_n)$ is the price vector, $\mathbf{a} = (a_1, \dots, a_k)$ is a vector of exogenous variables.

24.2 $\quad E_1(\mathbf{p}, \mathbf{a}) = 0, \; E_2(\mathbf{p}, \mathbf{a}) = 0, \; \dots, \; E_n(\mathbf{p}, \mathbf{a}) = 0$

Conditions for equilibrium.

24.3
$$E_1(p_1, p_2, a_1, \dots, a_k) = 0$$
$$E_2(p_1, p_2, a_1, \dots, a_k) = 0$$

Equilibrium conditions for the two good case.

24.4
$$\frac{\partial p_1}{\partial a_j} = \frac{\dfrac{\partial E_1}{\partial p_2}\dfrac{\partial E_2}{\partial a_j} - \dfrac{\partial E_2}{\partial p_2}\dfrac{\partial E_1}{\partial a_j}}{\dfrac{\partial E_1}{\partial p_1}\dfrac{\partial E_2}{\partial p_2} - \dfrac{\partial E_1}{\partial p_2}\dfrac{\partial E_2}{\partial p_1}}$$

$$\frac{\partial p_2}{\partial a_j} = \frac{\dfrac{\partial E_2}{\partial p_1}\dfrac{\partial E_1}{\partial a_j} - \dfrac{\partial E_1}{\partial p_1}\dfrac{\partial E_2}{\partial a_j}}{\dfrac{\partial E_1}{\partial p_1}\dfrac{\partial E_2}{\partial p_2} - \dfrac{\partial E_1}{\partial p_2}\dfrac{\partial E_2}{\partial p_1}}$$

Comparative statics results for the two good case, $j = 1, \dots, k$.

24.5
$$\begin{pmatrix} \dfrac{\partial p_1}{\partial a_j} \\ \vdots \\ \dfrac{\partial p_n}{\partial a_j} \end{pmatrix} = - \begin{pmatrix} \dfrac{\partial E_1}{\partial p_1} & \cdots & \dfrac{\partial E_1}{\partial p_n} \\ \vdots & \ddots & \vdots \\ \dfrac{\partial E_n}{\partial p_1} & \cdots & \dfrac{\partial E_n}{\partial p_n} \end{pmatrix}^{-1} \begin{pmatrix} \dfrac{\partial E_1}{\partial a_j} \\ \vdots \\ \dfrac{\partial E_n}{\partial a_j} \end{pmatrix}$$

Comparative statics results for the n good case, $j = 1, \dots, k$. See (19.16) for the general formula for the inverse of a square matrix.

Consider the problem

$$\max f(\mathbf{x}, \mathbf{a}) \text{ subject to } g(\mathbf{x}, \mathbf{a}) = 0$$

24.6

where f and g are C^1 functions, and let $\mathcal{L}$ be the associated Lagrangian function, with Lagrange multiplier λ. If $x_i^* = x_i^*(\mathbf{a})$, $i = 1, \ldots, n$, solves the problem, then for $i, j = 1, \ldots, m$,

$$\sum_{k=1}^{n} L''_{a_i x_k} \frac{\partial x_k^*}{\partial a_j} + g'_{a_i} \frac{\partial \lambda}{\partial a_j} = \sum_{k=1}^{n} L''_{a_j x_k} \frac{\partial x_k^*}{\partial a_i} + g'_{a_j} \frac{\partial \lambda}{\partial a_i}$$

Reciprocity relations. $\mathbf{x} = (x_1, \ldots, x_n)$ are the decision variables, $\mathbf{a} = (a_1, \ldots, a_m)$ are the parameters. For a systematic use of these relations, see Silberberg (1990).

Monotone comparative statics

24.7

A function $F : Z \to \mathbb{R}$, defined on a sublattice Z of $\mathbb{R}^m$, is called *supermodular* if

$$F(\mathbf{z}) + F(\mathbf{z}') \leq F(\mathbf{z} \wedge \mathbf{z}') + F(\mathbf{z} \vee \mathbf{z}')$$

for all $\mathbf{z}$ and $\mathbf{z}'$ in Z. If the inequality is strict whenever $\mathbf{z}$ and $\mathbf{z}'$ are not comparable under the preordering $\leq$, then F is called *strictly supermodular*.

Definition of (strict) supermodularity. See (6.30) and (6.31) for the definition of a sublattice and the lattice operations $\wedge$ and $\vee$.

24.8

Let S and P be sublattices of $\mathbb{R}^n$ and $\mathbb{R}^l$, respectively. A function $f : S \times P \to \mathbb{R}$ is said to satisfy *increasing differences* in $(\mathbf{x}, \mathbf{p})$ if

$$\mathbf{x} \geq \mathbf{x}' \text{ and } \mathbf{p} \geq \mathbf{p}' \Rightarrow$$
$$f(\mathbf{x}, \mathbf{p}) - f(\mathbf{x}', \mathbf{p}) \geq f(\mathbf{x}, \mathbf{p}') - f(\mathbf{x}', \mathbf{p}')$$

for all pairs $(\mathbf{x}, \mathbf{p})$ and $(\mathbf{x}', \mathbf{p}')$ in $S \times P$. If the inequality is strict whenever $\mathbf{x} > \mathbf{x}'$ and $\mathbf{p} > \mathbf{p}'$, then f is said to satisfy *strictly increasing differences in* $(\mathbf{x}, \mathbf{p})$.

Definition of (strictly) increasing differences. (The difference $f(\mathbf{x}, \mathbf{p}) - f(\mathbf{x}', \mathbf{p})$ between the values of f evaluated at the larger "action" $\mathbf{x}$ and the lesser "action" $\mathbf{x}'$ is a (strictly) increasing function of the parameter $\mathbf{p}$.)

24.9

Let S and P be sublattices of $\mathbb{R}^n$ and $\mathbb{R}^l$, respectively. If $f : S \times P \to \mathbb{R}$ is supermodular in $(\mathbf{x}, \mathbf{p})$, then

- f is supermodular in $\mathbf{x}$ for fixed $\mathbf{p}$, i.e. for every fixed $\mathbf{p}$ in P, and for all $\mathbf{x}$ and $\mathbf{x}'$ in S,

$$f(\mathbf{x}, \mathbf{p}) + f(\mathbf{x}', \mathbf{p}) \leq f(\mathbf{x} \wedge \mathbf{x}', \mathbf{p}) + f(\mathbf{x} \vee \mathbf{x}', \mathbf{p});$$

- f satisfies increasing differences in $(\mathbf{x}, \mathbf{p})$.

Important facts. Note that $S \times P$ is a sublattice of $\mathbb{R}^n \times \mathbb{R}^l = \mathbb{R}^{n+l}$.

24.10
Let X be an open sublattice of $\mathbb{R}^m$. A C^2 function $F : X \to \mathbb{R}$ is supermodular on X if and only if for all $\mathbf{x}$ in X,

$$\frac{\partial^2 F}{\partial x_i \partial x_j}(\mathbf{x}) \geq 0, \quad i, j = 1, \ldots, m, \quad i \neq j$$

24.11
Suppose that the problem

$$\max F(x, p) \quad \text{subject to} \quad x \in S \subset \mathbb{R}$$

has at least one solution for each $p \in P \subset \mathbb{R}$. Suppose in addition that F satisfies strictly increasing differences in (x, p). Then the optimal action $x^*(p)$ is increasing in the parameter p.

> A special result that cannot be extended to the case $S \subset \mathbb{R}^n$ for $n \geq 2$.

24.12
Suppose in (24.11) that

$$F(x, p) = pf(x) - C(x)$$

with S compact and f and C continuous. Then $\partial^2 F / \partial x \partial p = f'(x)$, so according to (24.10), F is supermodular if and only if $f(x)$ is increasing. Thus $f(x)$ increasing is sufficient to ensure that the optimal action $x^*(p)$ is increasing in p.

> An important consequence of (24.10).

24.13
Suppose S is a compact sublattice of $\mathbb{R}^n$ and P a sublattice of $\mathbb{R}^l$ and $f : S \times P \to \mathbb{R}$ is a continuous function on S for each fixed $\mathbf{p}$. Suppose that f satisfies increasing differences in $(\mathbf{x}, \mathbf{p})$, and is supermodular in $\mathbf{x}$ for each fixed $\mathbf{p}$. Let the correspondence Γ from P to S be defined by

$$\Gamma(\mathbf{p}) = \operatorname{argmax}\{f(\mathbf{x}, \mathbf{p}) : \mathbf{x} \in S\}$$

- For each $\mathbf{p}$ in P, $\Gamma(\mathbf{p})$ is a nonempty compact sublattice of $\mathbb{R}^n$, and has a greatest element, denoted by $\mathbf{x}^*(\mathbf{p})$.
- $\mathbf{p}_1 > \mathbf{p}_2 \Rightarrow \mathbf{x}^*(\mathbf{p}_1) \geq \mathbf{x}^*(\mathbf{p}_2)$
- If f satisfies strictly increasing differences in $(\mathbf{x}, \mathbf{p})$, then $\mathbf{x}_1 \geq \mathbf{x}_2$ for all $\mathbf{x}_1$ in $\Gamma(\mathbf{p}_1)$ and all $\mathbf{x}_2$ in $\Gamma(\mathbf{p}_2)$ whenever $\mathbf{p}_1 > \mathbf{p}_2$.

> A main result. For a given $\mathbf{p}$, $\operatorname{argmax}\{f(\mathbf{x}, \mathbf{p}) : \mathbf{x} \in S\}$ is the set of all points $\mathbf{x}$ in S where $f(\mathbf{x}, \mathbf{p})$ attains its maximum value.

References

On comparative statics, see Varian (1992) or Silberberg (1990). On monotone comparative statics, see Sundaram (1996) and Topkis (1998).

Chapter 25

Properties of cost and profit functions

25.1 $\quad C(\mathbf{w}, y) = \min\limits_{\mathbf{x}} \sum\limits_{i=1}^{n} w_i x_i \quad$ when $\quad f(\mathbf{x}) = y$

Cost minimization. One output. f is the production function, $\mathbf{w} = (w_1, \ldots, w_n)$ are factor prices, y is output and $\mathbf{x} = (x_1, \ldots, x_n)$ are factor inputs. $C(\mathbf{w}, y)$ is the cost function.

25.2 $\quad C(\mathbf{w}, y) = \begin{cases} \text{The minimum cost of producing} \\ y \text{ units of a commodity when fac-} \\ \text{tor prices are } \mathbf{w} = (w_1, \ldots, w_n). \end{cases}$

The cost function.

25.3
- $C(\mathbf{w}, y)$ is increasing in each w_i.
- $C(\mathbf{w}, y)$ is homogeneous of degree 1 in $\mathbf{w}$.
- $C(\mathbf{w}, y)$ is concave in $\mathbf{w}$.
- $C(\mathbf{w}, y)$ is continuous in $\mathbf{w}$ for $\mathbf{w} > \mathbf{0}$.

Properties of the cost function.

25.4 $\quad x_i^*(\mathbf{w}, y) = \begin{cases} \text{The cost minimizing choice of} \\ \text{the } i\text{th input factor as a func-} \\ \text{tion of the factor prices } \mathbf{w} \text{ and} \\ \text{the production level } y. \end{cases}$

Conditional factor demand functions. $\mathbf{x}^*(\mathbf{w}, y)$ is the vector $\mathbf{x}^*$ that solves the problem in (25.1).

25.5
- $x_i^*(\mathbf{w}, y)$ is decreasing in w_i.
- $x_i^*(\mathbf{w}, y)$ is homogeneous of degree 0 in $\mathbf{w}$.

Properties of the conditional factor demand function.

25.6 $\quad \dfrac{\partial C(\mathbf{w}, y)}{\partial w_i} = x_i^*(\mathbf{w}, y), \quad i = 1, \ldots, n$

Shephard's lemma.

25.7 $\quad \left(\dfrac{\partial^2 C(\mathbf{w}, y)}{\partial w_i \partial w_j} \right)_{(n \times n)} = \left(\dfrac{\partial x_i^*(\mathbf{w}, y)}{\partial w_j} \right)_{(n \times n)}$
is symmetric and negative semidefinite.

Properties of the substitution matrix.

25.8 $\quad \pi(p, \mathbf{w}) = \max_{\mathbf{x}} \left(p f(\mathbf{x}) - \sum_{i=1}^{n} w_i x_i \right)$

The profit maximizing problem of the firm. p is the price of output. $\pi(p, \mathbf{w})$ is the *profit function*.

25.9 $\quad \pi(p, \mathbf{w}) = \begin{cases} \text{The maximum profit as a function} \\ \text{of the factor prices } \mathbf{w} \text{ and the out-} \\ \text{put price } p. \end{cases}$

The profit function.

25.10 $\quad \pi(p, \mathbf{w}) \equiv \max_{y} \left(py - C(\mathbf{w}, y) \right)$

The profit function in terms of costs and revenue.

25.11
- $\pi(p, \mathbf{w})$ is increasing in p.
- $\pi(p, \mathbf{w})$ is homogeneous of degree 1 in $(p, \mathbf{w})$.
- $\pi(p, \mathbf{w})$ is convex in $(p, \mathbf{w})$.
- $\pi(p, \mathbf{w})$ is continuous in $(p, \mathbf{w})$ for $\mathbf{w} > \mathbf{0}$, $p > 0$.

Properties of the profit function.

25.12 $\quad x_i(p, \mathbf{w}) = \begin{cases} \text{The profit maximizing choice of} \\ \text{the } i\text{th input factor as a function} \\ \text{of the price of output } p \text{ and the} \\ \text{factor prices } \mathbf{w}. \end{cases}$

The *factor demand functions*. $\mathbf{x}(p, \mathbf{w})$ is the vector $\mathbf{x}$ that solves the problem in (25.8).

25.13
- $x_i(p, \mathbf{w})$ is decreasing in w_i.
- $x_i(p, \mathbf{w})$ is homogeneous of degree 0 in $(p, \mathbf{w})$.
  The *cross-price effects* are symmetric:
  $$\frac{\partial x_i(p, \mathbf{w})}{\partial w_j} = \frac{\partial x_j(p, \mathbf{w})}{\partial w_i}, \quad i, j = 1, \ldots, n$$

Properties of the factor demand functions.

25.14 $\quad y(p, \mathbf{w}) = \begin{cases} \text{The profit maximizing output as} \\ \text{a function of the price of output } p \\ \text{and the factor prices } \mathbf{w}. \end{cases}$

The *supply function* $y(p, \mathbf{w}) = f(\mathbf{x}(p, \mathbf{w}))$ is the y that solves the problem in (25.10).

25.15
- $y(p, \mathbf{w})$ is increasing in p.
- $y(p, \mathbf{w})$ is homogeneous of degree 0 in $(p, \mathbf{w})$.

Properties of the supply function.

25.16 $\quad \dfrac{\partial \pi(p, \mathbf{w})}{\partial p} = y(p, \mathbf{w})$

$\dfrac{\partial \pi(p, \mathbf{w})}{\partial w_i} = -x_i(p, \mathbf{w}), \quad i = 1, \ldots, n$

Hotelling's lemma.

$$25.17 \quad \frac{\partial x_j(p, \mathbf{w})}{\partial w_k} = \frac{\partial x_j^*(\mathbf{w}, y)}{\partial w_k} + \frac{\dfrac{\partial x_j(p, \mathbf{w})}{\partial p} \dfrac{\partial y(p, \mathbf{w})}{\partial w_k}}{\dfrac{\partial y(p, \mathbf{w})}{\partial p}}$$

Puu's equation, $j, k = 1, \ldots, n$, shows the substitution and scale effects of an increase in a factor price.

Elasticities of substitution in production theory

$$25.18 \quad \sigma_{yx} = \mathrm{El}_{R_{yx}} \left(\frac{y}{x} \right) = -\frac{\partial \ln \left(\dfrac{y}{x} \right)}{\partial \ln \left(\dfrac{p_2}{p_1} \right)}, \quad f(x, y) = c$$

The *elasticity of substitution* between y and x, assuming factor markets are competitive. (See also (5.20).)

$$25.19 \quad \sigma_{ij} = -\frac{\partial \ln \left(\dfrac{C_i'(\mathbf{w}, y)}{C_j'(\mathbf{w}, y)} \right)}{\partial \ln \left(\dfrac{w_i}{w_j} \right)}, \quad i \neq j$$

y, C, and w_k (for $k \neq i, j$) are constants.

The *shadow elasticity of substitution* between factor i and factor j.

$$25.20 \quad \sigma_{ij} = \frac{-\dfrac{C_{ii}''}{(C_i')^2} + \dfrac{2C_{ij}''}{C_i' C_j'} - \dfrac{C_{jj}''}{(C_j')^2}}{\dfrac{1}{w_i C_i'} + \dfrac{1}{w_j C_j'}}, \quad i \neq j$$

An alternative form of (25.19).

$$25.21 \quad A_{ij}(\mathbf{w}, y) = \frac{C(\mathbf{w}, y) C_{ij}''(\mathbf{w}, y)}{C_i'(\mathbf{w}, y) C_j'(\mathbf{w}, y)}, \quad i \neq j$$

The *Allen–Uzawa elasticity of substitution.*

$$25.22 \quad A_{ij}(\mathbf{w}, y) = \frac{\varepsilon_{ij}(\mathbf{w}, y)}{S_j(\mathbf{w}, y)}, \quad i \neq j$$

Here $\varepsilon_{ij}(\mathbf{w}, y)$ is the (constant-output) cross-price elasticity of demand, and $S_j(\mathbf{w}, y) = p_j C_j(\mathbf{w}, y)/C(\mathbf{w}, y)$ is the share of the jth input in total cost.

$$25.23 \quad M_{ij}(\mathbf{w}, y) = \frac{w_i C_{ij}''(\mathbf{w}, y)}{C_j'(\mathbf{w}, y)} - \frac{w_i C_{ii}''(\mathbf{w}, y)}{C_i'(\mathbf{w}, y)}$$
$$= \varepsilon_{ji}(\mathbf{w}, y) - \varepsilon_{ii}(\mathbf{w}, y), \quad i \neq j$$

The *Morishima elasticity of substitution.*

25.24 If $n > 2$, then $M_{ij}(\mathbf{w}, y) = M_{ji}(\mathbf{w}, y)$ for all $i \neq j$ if and only if all the $M_{ij}(\mathbf{w}, y)$ are equal to one and the same constant.

Symmetry of the Morishima elasticity of substitution.

Special functional forms and their properties

The Cobb–Douglas function

25.25 $\quad y = x_1^{a_1} x_2^{a_2} \cdots x_n^{a_n}$

> The Cobb–Douglas function, defined for $x_i > 0$, $i = 1, \ldots, n$. $a_1, \ldots, a_n$ are positive constants.

The Cobb–Douglas function in (25.25) is:

25.26
(a) homogeneous of degree $a_1 + \cdots + a_n$,
(b) quasiconcave for all $a_1, \ldots, a_n$,
(c) concave if $a_1 + \cdots + a_n \leq 1$,
(d) strictly concave if $a_1 + \cdots + a_n < 1$.

> Properties of the Cobb–Douglas function. ($a_1, \ldots, a_n$ are positive constants.)

25.27 $\quad x_k^*(\mathbf{w}, y) = \left(\dfrac{a_k}{w_k}\right)\left(\dfrac{w_1}{a_1}\right)^{\frac{a_1}{s}} \cdots \left(\dfrac{w_n}{a_n}\right)^{\frac{a_n}{s}} y^{\frac{1}{s}}$

> Conditional factor demand functions with $s = a_1 + \cdots + a_n$.

25.28 $\quad C(\mathbf{w}, y) = s\left(\dfrac{w_1}{a_1}\right)^{\frac{a_1}{s}} \cdots \left(\dfrac{w_n}{a_n}\right)^{\frac{a_n}{s}} y^{\frac{1}{s}}$

> The cost function with $s = a_1 + \cdots + a_n$.

25.29 $\quad \dfrac{w_k x_k^*}{C(\mathbf{w}, y)} = \dfrac{a_k}{a_1 + \cdots + a_n}$

> Factor shares in total costs.

25.30 $\quad x_k(p, \mathbf{w}) = \dfrac{a_k}{w_k}(pA)^{\frac{1}{1-s}}\left(\dfrac{w_1}{a_1}\right)^{\frac{a_1}{s-1}} \cdots \left(\dfrac{w_n}{a_n}\right)^{\frac{a_n}{s-1}}$

> Factor demand functions with $s = a_1 + \cdots + a_n < 1$.

25.31 $\quad \pi(p, \mathbf{w}) = (1 - s)(p)^{\frac{1}{1-s}} \displaystyle\prod_{i=1}^{n}\left(\dfrac{w_i}{a_i}\right)^{-\frac{a_i}{1-s}}$

> The profit function with $s = a_1 + \cdots + a_n < 1$. (If $s = a_1 + \cdots + a_n \geq 1$, there are increasing returns to scale, and the profit maximization problem has no solution.)

The CES (constant elasticity of substitution) function

25.32 $\quad y = (\delta_1 x_1^{-\rho} + \delta_2 x_2^{-\rho} + \cdots + \delta_n x_n^{-\rho})^{-\mu/\rho}$

> The CES function, defined for $x_i > 0$, $i = 1, \ldots, n$. μ and $\delta_1, \ldots, \delta_n$ are positive, and $\rho \neq 0$.

The CES function in (25.32) is:

25.33
(a) homogeneous of degree μ
(b) quasiconcave for $\rho \geq -1$,
   quasiconvex for $\rho \leq -1$
(c) concave for $\mu \leq 1$, $\rho \geq -1$
(d) convex for $\mu \geq 1$, $\rho \leq -1$

Properties of the CES function.

25.34 $\quad x_k^*(\mathbf{w}, y) = \dfrac{y^{\frac{1}{\mu}} w_k^{r-1}}{a_k^r} \left[\left(\dfrac{w_1}{a_1} \right)^r + \cdots + \left(\dfrac{w_n}{a_n} \right)^r \right]^{\frac{1}{\rho}}$

Conditional factor demand functions with $r = \rho/(\rho + 1)$ and $a_k = \delta_k^{-1/\rho}$.

25.35 $\quad C(\mathbf{w}, y) = y^{\frac{1}{\mu}} \left[\left(\dfrac{w_1}{a_1} \right)^r + \cdots + \left(\dfrac{w_n}{a_n} \right)^r \right]^{\frac{1}{r}}$

The cost function.

25.36 $\quad \dfrac{w_k x_k^*}{C(\mathbf{w}, y)} = \dfrac{\left(\dfrac{w_k}{a_k} \right)^r}{\left(\dfrac{w_1}{a_1} \right)^r + \cdots + \left(\dfrac{w_n}{a_n} \right)^r}$

Factor shares in total costs.

Law of the minimum

25.37 $\quad y = \min(a_1 + b_1 x_1, \ldots, a_n + b_n x_n)$

Law of the minimum. When $a_1 = \cdots = a_n = 0$, this is the *Leontief* or *fixed coefficient function*.

25.38 $\quad x_k^*(\mathbf{w}, y) = \dfrac{y - a_k}{b_k}, \quad k = 1, \ldots, n$

Conditional factor demand functions.

25.39 $\quad C(\mathbf{w}, y) = \left(\dfrac{y - a_1}{b_1} \right) w_1 + \cdots + \left(\dfrac{y - a_n}{b_n} \right) w_n$

The cost function.

The Diewert (generalized Leontief) cost function

25.40 $\quad C(\mathbf{w}, y) = y \sum_{i,j=1}^{n} b_{ij} \sqrt{w_i w_j} \quad$ with $b_{ij} = b_{ji}$

The *Diewert cost function*.

25.41 $\quad x_k^*(\mathbf{w}, y) = y \sum_{j=1}^{n} b_{kj} \sqrt{w_k / w_j}$

Conditional factor demand functions.

The translog cost function

$$\ln C(\mathbf{w}, y) = a_0 + c_1 \ln y + \sum_{i=1}^{n} a_i \ln w_i$$

25.42
$$+ \frac{1}{2} \sum_{i,j=1}^{n} a_{ij} \ln w_i \ln w_j + \sum_{i=1}^{n} b_i \ln w_i \ln y$$

Restrictions: $\sum_{i=1}^{n} a_i = 1$, $\sum_{i=1}^{n} b_i = 0$,

$$\sum_{j=1}^{n} a_{ij} = \sum_{i=1}^{n} a_{ij} = 0, \qquad i,j = 1, \ldots, n$$

The translog cost function. $a_{ij} = a_{ji}$ for all i and j. The restrictions on the coefficients ensure that $C(\mathbf{w}, y)$ is homogeneous of degree 1.

25.43
$$\frac{w_k x_k^*}{C(\mathbf{w}, y)} = a_k + \sum_{j=1}^{n} a_{kj} \ln w_j + b_i \ln y$$

Factor shares in total costs.

References

Varian (1992) is a basic reference. For a detailed discussion of existence and differentiability assumptions, see Fuss and McFadden (1978). For a discussion of Puu's equation (25.17), see Johansen (1972). For (25.18)–(25.24), see Blackorby and Russell (1989). For special functional forms, see Fuss and McFadden (1978).

Chapter 26

Consumer theory

26.1 A *preference relation* $\succeq$ on a set X of commodity vectors $\mathbf{x} = (x_1, \ldots, x_n)$ is a complete, reflexive, and transitive binary relation on X with the interpretation

$\mathbf{x} \succeq \mathbf{y}$ means: $\mathbf{x}$ is at least as good as $\mathbf{y}$

Definition of a preference relation. For binary relations, see (1.16).

26.2 Relations derived from $\succeq$:

- $\mathbf{x} \sim \mathbf{y} \iff \mathbf{x} \succeq \mathbf{y}$ and $\mathbf{y} \succeq \mathbf{x}$
- $\mathbf{x} \succ \mathbf{y} \iff \mathbf{x} \succeq \mathbf{y}$ but not $\mathbf{y} \succeq \mathbf{x}$

$\mathbf{x} \sim \mathbf{y}$ is read "$\mathbf{x}$ is *indifferent* to $\mathbf{y}$", and $\mathbf{x} \succ \mathbf{y}$ is read "$\mathbf{x}$ is *(strictly) preferred* to $\mathbf{y}$".

26.3
- A function $u : X \to \mathbb{R}$ is a *utility function representing the preference relation* $\succeq$ if

$\mathbf{x} \succeq \mathbf{y} \iff u(\mathbf{x}) \geq u(\mathbf{y})$

- For any strictly increasing function $f : \mathbb{R} \to \mathbb{R}$, $u^*(\mathbf{x}) = f(u(\mathbf{x}))$ is a new utility function representing the same preferences as $u(\cdot)$.

A property of utility functions that is invariant under every strictly increasing transformation, is called *ordinal*. *Cardinal* properties are those *not* preserved under strictly increasing transformations.

26.4 Let $\succeq$ be a complete, reflexive, and transitive preference relation that is also *continuous* in the sense that the sets

$\{\mathbf{x} : \mathbf{x} \succeq \mathbf{x}^0\}$ and $\{\mathbf{x} : \mathbf{x}^0 \succeq \mathbf{x}\}$

are both closed for all $\mathbf{x}^0$ in X. Then $\succeq$ can be represented by a continuous utility function.

Existence of a continuous utility function. For properties of relations, see (1.16).

26.5 *Utility maximization* subject to a budget constraint:

$$\max_{\mathbf{x}} u(\mathbf{x}) \text{ subject to } \mathbf{p} \cdot \mathbf{x} = \sum_{i=1}^{n} p_i x_i = m$$

$\mathbf{x} = (x_1, \ldots, x_n)$ is a vector of (quantities of) commodities, $\mathbf{p} = (p_1, \ldots, p_n)$ is the price vector, m is income, and u is the utility function.

26.6 $v(\mathbf{p}, m) = \max_{\mathbf{x}}\{u(\mathbf{x}) : \mathbf{p} \cdot \mathbf{x} = m\}$

The *indirect utility function*, $v(\mathbf{p}, m)$, is the maximum utility as a function of the price vector $\mathbf{p}$ and the income m.

26.7

- $v(\mathbf{p}, m)$ is decreasing in $\mathbf{p}$.
- $v(\mathbf{p}, m)$ is increasing in m.
- $v(\mathbf{p}, m)$ is homogeneous of degree 0 in $(\mathbf{p}, m)$.
- $v(\mathbf{p}, m)$ is quasi-convex in $\mathbf{p}$.
- $v(\mathbf{p}, m)$ is continuous in $(\mathbf{p}, m)$, $\mathbf{p} > \mathbf{0}$, $m > 0$.

Properties of the indirect utility function.

26.8 $\omega = \dfrac{u'_1(\mathbf{x})}{p_1} = \cdots = \dfrac{u'_n(\mathbf{x})}{p_n}$

First-order conditions for problem (26.5), with ω as the associated Lagrange multiplier.

26.9 $\omega = \dfrac{\partial v(\mathbf{p}, m)}{\partial m}$

ω is called the *marginal utility* of money.

26.10 $x_i(\mathbf{p}, m) = \begin{cases} \text{the optimal choice of the } i\text{th com-} \\ \text{modity as a function of the price} \\ \text{vector } \mathbf{p} \text{ and the income } m. \end{cases}$

The *consumer demand functions*, or *Marshallian demand functions*, derived from problem (26.5).

26.11 $\mathbf{x}(t\mathbf{p}, tm) = \mathbf{x}(\mathbf{p}, m)$, t is a positive scalar.

The demand functions are homogeneous of degree 0.

26.12 $x_i(\mathbf{p}, m) = -\dfrac{\dfrac{\partial v(\mathbf{p}, m)}{\partial p_i}}{\dfrac{\partial v(\mathbf{p}, m)}{\partial m}}$, $i = 1, \ldots, n$

Roy's identity.

26.13 $e(\mathbf{p}, u) = \min_{\mathbf{x}}\{\mathbf{p} \cdot \mathbf{x} : u(\mathbf{x}) \geq u\}$

The *expenditure function*, $e(\mathbf{p}, u)$, is the minimum expenditure at prices $\mathbf{p}$ for obtaining at least the utility level u.

26.14

- $e(\mathbf{p}, u)$ is increasing in $\mathbf{p}$.
- $e(\mathbf{p}, u)$ is homogeneous of degree 1 in $\mathbf{p}$.
- $e(\mathbf{p}, u)$ is concave in $\mathbf{p}$.
- $e(\mathbf{p}, u)$ is continuous in $\mathbf{p}$ for $\mathbf{p} > \mathbf{0}$.

Properties of the expenditure function.

26.15 $\mathbf{h}(\mathbf{p}, u) = \begin{cases} \text{the expenditure-minimizing bun-} \\ \text{dle necessary to achieve utility} \\ \text{level } u \text{ at prices } \mathbf{p}. \end{cases}$

The *Hicksian (or compensated) demand function.* $\mathbf{h}(\mathbf{p}, u)$ is the vector $\mathbf{x}$ that solves the problem $\min_{\mathbf{x}}\{\mathbf{p} \cdot \mathbf{x} : u(\mathbf{x}) \geq u\}$.

26.16 $\dfrac{\partial e(\mathbf{p}, u)}{\partial p_i} = h_i(\mathbf{p}, u) \quad \text{for} \quad i = 1, \ldots, n$

Relationship between the expenditure function and the Hicksian demand function.

26.17 $\dfrac{\partial h_i(\mathbf{p}, u)}{\partial p_j} = \dfrac{\partial h_j(\mathbf{p}, u)}{\partial p_i}, \quad i, j = 1, \ldots, n$

Symmetry of the Hicksian cross partials. (The *Marshallian cross partials* need not be symmetric.)

26.18 The matrix $\mathbf{S} = (S_{ij})_{n \times n} = \left(\dfrac{\partial h_i(\mathbf{p}, u)}{\partial p_j}\right)_{n \times n}$ is negative semidefinite.

Follows from (26.16) and the concavity of the expenditure function.

26.19 $e(\mathbf{p}, v(\mathbf{p}, m)) = m : \begin{cases} \text{the minimum expenditure} \\ \text{needed to achieve utility} \\ v(\mathbf{p}, m) \text{ is } m. \end{cases}$

Useful identities that are valid except in rather special cases.

26.20 $v(\mathbf{p}, e(\mathbf{p}, u)) = u : \begin{cases} \text{the maximum utility from} \\ \text{income } e(\mathbf{p}, u) \text{ is } u. \end{cases}$

26.21 Marshallian demand at income m is Hicksian demand at utility $v(\mathbf{p}, m)$:
$$x_i(\mathbf{p}, m) = h_i(\mathbf{p}, v(\mathbf{p}, m))$$

26.22 Hicksian demand at utility u is the same as Marshallian demand at income $e(\mathbf{p}, u)$:
$$h_i(\mathbf{p}, u) = x_i(\mathbf{p}, e(\mathbf{p}, u))$$

26.23
- $e_{ij} = \mathrm{El}_{p_j}\, x_i = \dfrac{p_j}{x_i}\dfrac{\partial x_i}{\partial p_j}$ *(Cournot elasticities)*

- $E_i = \mathrm{El}_m\, x_i = \dfrac{m}{x_i}\dfrac{\partial x_i}{\partial m}$ *(Engel elasticities)*

- $S_{ij} = \mathrm{El}_{p_j}\, h_i = \dfrac{p_j}{x_i}\dfrac{\partial h_i}{\partial p_j}$ *(Slutsky elasticities)*

e_{ij} are the elasticities of demand w.r.t. prices, E_i are the elasticities of demand w.r.t. income, and S_{ij} are the elasticities of the Hicksian demand w.r.t. prices.

26.24
- $\dfrac{\partial x_i(\mathbf{p}, m)}{\partial p_j} = \dfrac{\partial h_i(\mathbf{p}, u)}{\partial p_j} - x_j(\mathbf{p}, m)\dfrac{\partial x_i(\mathbf{p}, m)}{\partial m}$

- $S_{ij} = e_{ij} + a_j E_i, \quad a_j = p_j x_j / m$

Two equivalent forms of the *Slutsky equation.*

The following $\frac{1}{2}n(n+1)+1$ restrictions on the partial derivatives of the demand functions are linearly independent:

26.25

(a) $\displaystyle\sum_{i=1}^{n} p_i \frac{\partial x_i(\mathbf{p},m)}{\partial m} = 1$

(b) $\displaystyle\sum_{j=1}^{n} p_j \frac{\partial x_i}{\partial p_j} + m \frac{\partial x_i}{\partial m} = 0, \quad i = 1,\dots,n$

(c) $\dfrac{\partial x_i}{\partial p_j} + x_j \dfrac{\partial x_i}{\partial m} = \dfrac{\partial x_j}{\partial p_i} + x_i \dfrac{\partial x_j}{\partial m}$

for $1 \le i < j \le n$

(a) is the budget constraint differentiated with respect to m.
(b) is the Euler equation (for homogeneous functions) applied to the consumer demand function.
(c) is a consequence of the Slutsky equation and (26.17).

26.26

$EV = e(\mathbf{p}^0, v(\mathbf{p}^1, m^1)) - e(\mathbf{p}^0, v(\mathbf{p}^0, m^0))$

EV is the difference between the amount of money needed at the old (period 0) prices to reach the new (period 1) utility level, and the amount of money needed at the old prices to reach the old utility level.

Equivalent variation. $\mathbf{p}^0$, m^0, and $\mathbf{p}^1$, m^1, are prices and income in period 0 and period 1, respectively. $(e(\mathbf{p}^0, v(\mathbf{p}^0, m^0)) = m^0.)$

26.27

$CV = e(\mathbf{p}^1, v(\mathbf{p}^1, m^1)) - e(\mathbf{p}^1, v(\mathbf{p}^0, m^0))$

CV is the difference between the amount of money needed at the new (period 1) prices to reach the new utility level, and the amount of money needed at the new prices to reach the old (period 0) utility level.

Compensating variation. $\mathbf{p}^0$, m^0, and $\mathbf{p}^1$, m^1, are prices and income in period 0 and period 1, respectively. $(e(\mathbf{p}^1, v(\mathbf{p}^1, m^1)) = m^1.)$

Special functional forms and their properties

Linear expenditure system (LES)

26.28 $\displaystyle u(\mathbf{x}) = \prod_{i=1}^{n} (x_i - c_i)^{\beta_i}, \qquad \beta_i > 0$

The *Stone–Geary* utility function. If $c_i = 0$ for all i, $u(\mathbf{x})$ is Cobb–Douglas.

26.29 $\displaystyle x_i(\mathbf{p},m) = c_i + \frac{1}{p_i} \frac{\beta_i}{\beta} \left(m - \sum_{i=1}^{n} p_i c_i \right)$

The demand functions. $\beta = \sum_{i=1}^{n} \beta_i$.

26.30 $\displaystyle v(\mathbf{p},m) = \beta^{-\beta} \left(m - \sum_{i=1}^{n} p_i c_i \right)^{\beta} \prod_{i=1}^{n} \left(\frac{\beta_i}{p_i} \right)^{\beta_i}$

The indirect utility function.

$$26.31 \quad e(\mathbf{p}, u) = \sum_{i=1}^{n} p_i c_i + \frac{\beta u^{1/\beta}}{\left[\prod_{i=1}^{n} \left(\frac{\beta_i}{p_i} \right)^{\beta_i} \right]^{1/\beta}}$$

The expenditure function.

Almost ideal demand system (AIDS)

$$\ln(e(\mathbf{p}, u)) = a(\mathbf{p}) + u b(\mathbf{p}), \quad \text{where}$$

$$a(\mathbf{p}) = \alpha_0 + \sum_{i=1}^{n} \alpha_i \ln p_i + \frac{1}{2} \sum_{i,j=1}^{n} \gamma_{ij}^* \ln p_i \ln p_j$$

26.32

$$\text{and} \quad b(\mathbf{p}) = \beta_0 \prod_{i=1}^{n} p_i^{\beta_i}, \quad \text{with restrictions}$$

$$\sum_{i=1}^{n} \alpha_i = 1, \ \sum_{i=1}^{n} \beta_i = 0, \ \text{and} \ \sum_{i=1}^{n} \gamma_{ij}^* = \sum_{j=1}^{n} \gamma_{ij}^* = 0.$$

Almost ideal demand system, defined by the logarithm of the expenditure function. The restrictions make $e(\mathbf{p}, u)$ homogeneous of degree 1 in $\mathbf{p}$.

$$x_i(\mathbf{p}, m) = \frac{m}{p_i} \left(\alpha_i + \sum_{j=1}^{n} \gamma_{ij} \ln p_j + \beta_i \ln\left(\frac{m}{P}\right) \right),$$

where the price index P is given by

26.33

$$\ln P = \alpha_0 + \sum_{i=1}^{n} \alpha_i \ln p_i + \frac{1}{2} \sum_{i,j=1}^{n} \gamma_{ij} \ln p_i \ln p_j$$

with $\gamma_{ij} = \frac{1}{2}(\gamma_{ij}^* + \gamma_{ji}^*) = \gamma_{ji}$

The demand functions.

Translog indirect utility function

$$\ln v(\mathbf{p}, m) = \alpha_0 + \sum_{i=1}^{n} \alpha_i \ln\left(\frac{p_i}{m}\right) +$$

26.34

$$\frac{1}{2} \sum_{i,j=1}^{n} \beta_{ij}^* \ln\left(\frac{p_i}{m}\right) \ln\left(\frac{p_j}{m}\right)$$

The translog indirect utility function.

26.35

$$x_i(\mathbf{p}, m) = \frac{m}{p_i} \left(\frac{\alpha_i + \sum_{j=1}^{n} \beta_{ij} \ln(p_j/m)}{\sum_{i=1}^{n} \alpha_i + \sum_{i,j=1}^{n} \beta_{ij}^* \ln(p_i/m)} \right)$$

where $\beta_{ij} = \frac{1}{2}(\beta_{ij}^* + \beta_{ji}^*)$.

The demand functions.

Price indices

Consider a "basket" of n commodities. Define for $i = 1, \ldots, n$,

$q^{(i)}$ = number of units of good i in the basket

$p_0^{(i)}$ = price per unit of good i in year 0

26.36 $p_t^{(i)}$ = price per unit of good i in year t

A *price index*, P, for year t, with year 0 as the base year, is defined as

$$P = \frac{\sum_{i=1}^{n} p_t^{(i)} q^{(i)}}{\sum_{i=1}^{n} p_0^{(i)} q^{(i)}} \cdot 100$$

The most common definition of a price index. P is 100 times the cost of the basket in year t divided by the cost of the basket in year 0. (More generally, a (consumption) price index can be defined as any function $P(p_1, \ldots, p_n)$ of all the prices, homogeneous of degree 1 and increasing in each variable.)

26.37

- If the quantities $q^{(i)}$ in the formula for P are levels of consumption in the base year 0, P is called the *Laspeyres price index*.

- If the quantities $q^{(i)}$ are levels of consumption in the year t, P is called the *Paasche price index*.

Two important price indices.

26.38 $F = \sqrt{(\text{Laspeyres index}) \cdot (\text{Paasche index})}$

Fisher's ideal index.

References

Varian (1992) is a basic reference. For a more advanced treatment, see Mas-Colell, Whinston, and Green (1995). For AIDS, see Deaton and Muellbauer (1980), for translog, see Christensen, Jorgenson, and Lau (1975). See also Phlips (1983).

Chapter 27

Topics from trade theory

27.1 Standard neoclassical trade model (2×2 *factor model*). Two factors of production, K and L, that are mobile between two output producing sectors A and B. Production functions are neoclassical (i.e. the production set is closed, convex, contains zero, has free disposal, and its intersection with the positive orthant is empty) and exhibit constant returns to scale.

The economy has *incomplete specialization* when both goods are produced.

27.2 Good B is more K intensive than good A if $K_B/L_B > K_A/L_A$ at all factor prices.

No factor intensity reversal (NFIR). K_B denotes use of factor K in producing good B, etc.

27.3 *Stolper–Samuelson's theorem:*
In the 2×2 factor model with no factor intensity reversal and incomplete specialization, an increase in the relative price of a good results in an increase in the real return to the factor used intensively in producing that good and a fall in the real return to the other factor.

When B is more K intensive, an increase in the price of B leads to an increase in the real return to K and a decrease in the real return to L. With P as the price of output, r the return to K and w the return to L, r/P_A and r/P_B both rise while w/P_A and w/P_B both fall.

27.4 *Rybczynski's theorem:*
In a 2×2 factor model with no factor intensity reversal and incomplete specialization, if the endowment of a factor increases, the output of the good more intensive in that factor will increase while the output of the other good will fall.

Assumes that the endowment of the other factor does not change and that prices of outputs do not change, e.g. if K increases and B is K intensive, then the output of B will rise and the output of A will fall.

27.5

Heckscher–Ohlin–Samuelson model:

Two countries, two traded goods, two non-traded factors of production (K, L). The factors are in fixed supply in the two countries. The two countries have the same constant returns to scale production function for making B and A. Factor markets clear within each country and trade between the two countries clears the markets for the two goods. Each country has a zero balance of payments. Consumers in the two countries have identical homothetic preferences. There is perfect competition and there are no barriers to trade, including tariffs, transactions costs, or transport costs. Both countries' technologies exhibit no factor intensity reversals.

The HOS model.

27.6

Heckscher–Ohlin's theorem:

In the HOS model (27.5) with $K/L > K^*/L^*$ and with B being more K intensive at all factor prices, the home country exports good B.

The quantity version of the H–O model. A $*$ denotes foreign country values and the other country is referred to as the home country.

27.7

In the HOS model (27.5) with neither country specialized in the production of just one good, the price of K is the same in both countries and the price of L is the same in both countries.

Factor price equalization.

References

Mas-Colell, Whinston, and Green (1995) or Bhagwati, Panagariya, and Srinivasan (1998).

Chapter 28

Topics from finance and growth theory

| 28.1 | $S_t = S_{t-1} + rS_{t-1} = (1+r)S_{t-1}, \quad t = 1, 2, \ldots$ | In an account with interest rate r, an amount S_{t-1} increases after one period to S_t. |

| 28.2 | The *compound amount* S_t of a *principal* S_0 at the end of t periods at the interest rate r compounded at the end of each period is

 $S_t = S_0(1+r)^t$ | Compound interest. (The solution to the difference equation in (28.1).) |

| 28.3 | The amount S_0 that must be invested at the interest rate r compounded at the end of each period for t periods so that the compound amount will be S_t, is given by

 $S_0 = S_t(1+r)^{-t}$ | S_0 is called the *present value* of S_t. |

| 28.4 | When interest is compounded n times a year at regular intervals at the rate of r/n per period, then the effective annual interest is

 $\left(1 + \dfrac{r}{n}\right)^n - 1$ | *Effective annual rate of interest.* |

| 28.5 | $\begin{aligned} A_t &= \dfrac{R}{(1+r)^1} + \dfrac{R}{(1+r)^2} + \cdots + \dfrac{R}{(1+r)^t} \\ &= R\dfrac{1 - (1+r)^{-t}}{r} \end{aligned}$ | The *present value A_t* of an *annuity* of R per period for t periods at the interest rate of r per period. Payments at the end of each period. |

| 28.6 | The present value A of an annuity of R per period for an infinite number of periods at the interest rate of r per period, is

 $A = \dfrac{R}{(1+r)^1} + \dfrac{R}{(1+r)^2} + \cdots = \dfrac{R}{r}$ | The present value of an infinite annuity. First payment after one period. |

28.7 $\quad T = \dfrac{\ln\left(\dfrac{R}{R - rA}\right)}{\ln(1 + r)}$

The number T of periods needed to pay off a loan of A with periodic payment R and interest rate r per period.

28.8 $\quad S_t = (1 + r)S_{t-1} + (y_t - x_t), \quad t = 1, 2, \ldots$

In an account with interest rate r, an amount S_{t-1} increases after one period to S_t, if y_t are the deposits and x_t are the withdrawals in period t.

28.9 $\quad S_t = (1 + r)^t S_0 + \displaystyle\sum_{k=1}^{t} (1 + r)^{t-k}(y_k - x_k)$

The solution of equation (28.8)

28.10 $\quad S_t = (1 + r_t)S_{t-1} + (y_t - x_t), \quad t = 1, 2, \ldots$

Generalization of (28.8) to the case with a variable interest rate, r_t.

28.11 $\quad D_k = \dfrac{1}{\prod_{s=1}^{k}(1 + r_s)}$

The *discount factor* associated with (28.10). (Discounted from period k to period 0.)

28.12 $\quad R_k = \dfrac{D_k}{D_t} = \displaystyle\prod_{s=k+1}^{t} (1 + r_s)$

The *interest factor* associated with (28.10).

28.13 $\quad S_t = R_0 S_0 + \displaystyle\sum_{k=1}^{t} R_k(y_k - x_k)$

The solution of (28.10). R_k is defined in (28.12). (Generalizes (28.9).)

28.14 $\quad a_0 + \dfrac{a_1}{1 + r} + \dfrac{a_2}{(1 + r)^2} + \cdots + \dfrac{a_n}{(1 + r)^n} = 0$

r is the *internal rate of return* of an investment project. Negative a_t represents outlays, positive a_t represents receipts at time t.

28.15 If $a_0 < 0$ and $a_1, \ldots, a_n$ are all ≥ 0, then (28.14) has a unique solution $1 + r^* > 0$, i.e. a unique internal rate of return $r^* > -1$. The internal rate of return is positive provided $\sum_{i=0}^{n} a_i > 0$.

Consequence of Descartes's rule of signs (2.12).

28.16 $\quad A_0 = a_0, \; A_1 = a_0 + a_1, \; A_2 = a_0 + a_1 + a_2, \; \ldots,$
$A_n = a_0 + a_1 + \cdots + a_n$

The *accumulated cash flow* associated with (28.14).

28.17	If $A_n \neq 0$, and the sequence $A_0, A_1, \ldots, A_n$ changes sign only once, then (28.14) has a unique positive internal rate of return.	*Norstrøm's rule.*

28.18	The amount in an account after t years if K dollars earn continuous compound interest at the rate r is $$Ke^{rt}$$	*Continuous compound interest.*

28.19	The effective annual interest with continuous compounding at the interest rate r is $$e^r - 1$$	*Effective rate of interest*, with continuous compounding.

28.20	$Ke^{-rt}, \quad r = p/100$	The *present value* (with continuous compounding) of an amount K due in t years, if the interest is $p\%$ per year.

28.21	The discounted present value at time 0 of a continuous income stream at the rate $K(t)$ dollars per year over the time interval $[0, T]$, and with continuous compounding at the rate of interest r, is $$\int_0^T K(t)e^{-rt}\, dt$$	*Discounted present value*, continuous compounding.

28.22	The discounted present value at time s, of a continuous income stream at the rate $K(t)$ dollars per year over the time interval $[s, T]$, and with continuous compounding at the rate of interest r, is $$\int_s^T K(t)e^{-r(t-s)}\, dt$$	*Discounted present value*, continuous compounding.

28.23	*Solow's growth model:* • $X(t) = F\left(K(t), L(t)\right)$ • $\dot{K}(t) = sX(t)$ • $L(t) = L_0 e^{\lambda t}$	$X(t)$ is national income, $K(t)$ is capital, and $L(t)$ is the labor force at time t. F is a production function. s (the savings rate), λ, and L_0 are positive constants.

28.24	If F is homogeneous of degree 1, $k(t) = K(t)/L(t)$ is capital per worker, and $f(k) = F(k, 1)$, then (28.23) reduces to $$\dot{k} = sf(k) - \lambda k, \quad k(0) \text{ is given}$$	A simplified version of (28.23).

28.25 If $f(0) = 0$, $\lambda/s < f'(0) < \infty$, $f'(k) \to 0$ as $k \to \infty$, and $f''(k) \leq 0$ for all $k \geq 0$, then the equation in (28.24) has a unique solution on $[0, \infty)$ for every positive initial value $k(0)$ of k. The equation has a unique positive equilibrium state k^*, defined by $sf(k^*) = \lambda k^*$. This equilibrium is asymptotically stable on $(0, \infty)$. (See the figure below.)

Existence and uniqueness of solutions of Solow's growth model over $[0, \infty)$. (See Example 5.8.8 in Sydsæter et al. (2005).) The existence of k^* follows immediately from the conditions on f.

28.26

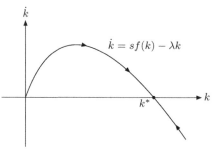

Phase diagram for (28.24), with the conditions in (28.25) imposed.

Ramsey's growth model:

28.27
$$\max \int_0^T U(C(t))e^{-rt}\,dt \quad \text{subject to}$$
$$C(t) = f(K(t)) - \dot{K}(t),$$
$$K(0) = K_0, \quad K(T) \geq K_1.$$

A standard problem in growth theory. U is a utility function, $K(t)$ is the capital stock at time t, $f(K)$ is the production function, $C(t)$ is consumption, r is the discount factor, and T is the planning horizon.

28.28 $$\ddot{K} - f'(K)\dot{K} + \frac{U'(C)}{U''(C)}(r - f'(K)) = 0$$

The Euler equation for problem (28.27).

28.29 $$\frac{\dot{C}}{C} = \frac{f'(K) - r}{-\check{w}}$$

where $\check{w} = \text{El}_C\, U'(C) = CU''(C)/U'(C)$

Necessary condition for the solution of (28.27). Since $\check{w}$ is usually negative, consumption increases if and only if the marginal productivity of capital exceeds the discount rate.

References

For compound interest formulas, see Goldberg (1961) or Sydsæter and Hammond (2005). For (28.17), see Norstrøm (1972). For growth theory, see Burmeister and Dobell (1970), Blanchard and Fischer (1989), Barro and Sala-i-Martin (1995), or Sydsæter et al. (2005).

Chapter 29

Risk and risk aversion theory

29.1 $$R_A = -\frac{u''(y)}{u'(y)}, \qquad R_R = yR_A = -\frac{yu''(y)}{u'(y)}$$

Absolute risk aversion (R_A) and *relative risk aversion* (R_R). $u(y)$ is a utility function, y is income, or consumption.

29.2
- $R_A = \lambda \iff u(y) = A_1 + A_2 e^{-\lambda y}$
- $R_R = k \iff u(y) = \begin{cases} A_1 + A_2 \ln y & \text{if } k = 1 \\ A_1 + A_2 y^{1-k} & \text{if } k \neq 1 \end{cases}$

A characterization of utility functions with constant absolute and relative risk aversion, respectively. A_1 and A_2 are constants, $A_2 \neq 0$.

29.3
- $u(y) = y - \frac{1}{2}by^2 \Rightarrow R_A = \dfrac{b}{1 - by}$
- $u(y) = \dfrac{1}{b-1}(a + by)^{1-\frac{1}{b}} \Rightarrow R_A = \dfrac{1}{a + by}$

Risk aversions for two special utility functions.

29.4 $$E[u(y + z + \pi)] = E[u(y)]$$
$$\pi \approx -\frac{u''(y)}{u'(y)}\frac{\sigma^2}{2} = R_A\frac{\sigma^2}{2}$$

Arrow–Pratt risk premium. π: risk premium. z: mean zero risky prospect. $\sigma^2 = \text{var}[z]$: variance of z. $E[\ ]$ is expectation. (Expectation and variance are defined in Chapter 33.)

29.5 If F and G are cumulative distribution functions (CDF) of random incomes, then

F first-degree stochastically dominates G
$$\iff G(Z) \geq F(Z) \text{ for all } Z \text{ in } I.$$

Definition of first-degree stochastic dominance. I is a closed interval $[Z_1, Z_2]$. For $Z \leq Z_1$, $F(Z) = G(Z) = 0$ and for $Z \geq Z_2$, $F(Z) = G(Z) = 1$.

29.6 F FSD $G \iff \begin{cases} E_F[u(Z)] \geq E_G[u(Z)] \\ \text{for all increasing } u(Z). \end{cases}$

An important result. FSD means "first-degree stochastically dominates". $E_F[u(Z)]$ is expected utility of income Z when the cumulative distribution function is $F(Z)$. $E_G[u(Z)]$ is defined similarly.

29.7 $T(Z) = \displaystyle\int_{Z_1}^{Z} (G(z) - F(z))\, dz$

A definition used in (29.8).

29.8 F *second-degree stochastically dominates* G

$\iff T(Z) \geq 0$ for all Z in I.

Definition of second-degree stochastic dominance (SSD). $I = [Z_1, Z_2]$. Note that FSD $\Rightarrow$ SSD.

29.9 F SSD $G \iff \begin{cases} E_F[u(Z)] \geq E_G[u(Z)] \text{ for all} \\ \text{increasing and concave } u(Z). \end{cases}$

Hadar–Russell's theorem. Every risk averter prefers F to G if and only if F SSD G.

29.10 Let F and G be distribution functions for X and Y, respectively, let $I = [Z_1, Z_2]$, and let $T(Z)$ be as defined in (29.7). Then the following statements are equivalent:

- $T(Z_2) = 0$ and $T(Z) \geq 0$ for all Z in I.
- There exists a stochastic variable ε with $E[\varepsilon \,|\, X] = 0$ for all X such that Y is distributed as $X + \varepsilon$.
- F and G have the same mean, and every risk averter prefers F to G.

Rothschild–Stiglitz's theorem.

References

See Huang and Litzenberger (1988), Hadar and Russell (1969), and Rothschild and Stiglitz (1970).

Chapter 30

Finance and stochastic calculus

Capital asset pricing model:

30.1

$$E[r_i] = r + \beta_i(E[r_m] - r)$$

where $\beta_i = \dfrac{\mathrm{corr}(r_i, r_m)\sigma_i}{\sigma_m} = \dfrac{\mathrm{cov}(r_i, r_m)}{\sigma_m^2}.$

r_i: rate of return on asset i. $E[r_k]$: the expected value of r_k. r: rate of return on a safe asset. r_m: market rate of return. σ_i: standard deviation of r_i.

Single consumption β asset pricing equation:

30.2

$$E(r_i) = r + \frac{\beta_{ic}}{\beta_{mc}}(E(r_m) - r),$$

where $\beta_{jc} = \dfrac{\mathrm{cov}(r_j, d\ln C)}{\mathrm{var}(d\ln C)}, \quad j = i$ or $m.$

C: consumption. r_m: return on an arbitrary portfolio. $d\ln C$ is the stochastic logarithmic differential. (See (30.13).)

The Black–Scholes option pricing model. (European or American call option on a stock that pays no dividend):

30.3

$$c = c(S, K, t, r, \sigma) = SN(x) - KN(x - \sigma\sqrt{t})e^{-rt},$$

where $x = \dfrac{\ln(S/K) + (r + \frac{1}{2}\sigma^2)t}{\sigma\sqrt{t}},$

and $N(y) = \frac{1}{\sqrt{2\pi}}\int_{-\infty}^{y} e^{-z^2/2}\, dz$ is the cumulative normal distribution function.

c: the value of the option on S at time t. S: underlying stock price, $dS/S = \alpha\, dt + \sigma\, dB$, where B is a (standard) Brownian motion, α: drift parameter. σ: *volatility* (measures the deviation from the mean). t: time left until expiration. r: interest rate. K: strike price.

30.4

- $\partial c/\partial S = N(x) > 0$
- $\partial c/\partial K = -N(x - \sigma\sqrt{t})e^{-rt} < 0$
- $\partial c/\partial t = \dfrac{\sigma}{2\sqrt{t}}SN'(x) + re^{-rt}KN(x - \sigma\sqrt{t}) > 0$
- $\partial c/\partial r = tKN(x - \sigma\sqrt{t})e^{-rt} > 0$
- $\partial c/\partial\sigma = SN'(x)\sqrt{t} > 0$

Useful sensitivity results for the Black–Scholes model. (The corresponding results for the generalized Black–Scholes model (30.5) are given in Haug (1997), Appendix B.)

The *generalized Black–Scholes model*, which includes the *cost-of-carry term b* (used to price European call options (c) and put options (p) on assets paying a continuous dividend yield, options on futures, and currency options):

30.5

$$c = SN(x)e^{(b-r)t} - KN(x - \sigma\sqrt{t})e^{-rt},$$

$$p = KN(\sigma\sqrt{t} - x)e^{-rt} - SN(-x)e^{(b-r)t},$$

where $x = \dfrac{\ln(S/K) + (b + \frac{1}{2}\sigma^2)t}{\sigma\sqrt{t}}$.

b: cost-of-carry rate of holding the underlying security. $b = r$ gives the Black–Scholes model. $b = r - q$ gives the Merton stock option model with continuous dividend yield q. $b = 0$ gives the Black futures option model.

30.6 $p = c - Se^{(b-r)t} + Ke^{-rt}$

The *put-call parity* for the generalized Black–Scholes model.

30.7 $P(S, K, t, r, b, \sigma) = C(K, S, t, r - b, -b, \sigma)$

A transformation that gives the formula for an American put option, P, in terms of the corresponding call option, C.

The market value of an *American perpetual put option* when the underlying asset pays no dividend:

30.8

$$h(x) = \begin{cases} \dfrac{K}{1+\gamma}\left(\dfrac{x}{c}\right)^{-\gamma} & \text{if } x \geq c, \\ K - x & \text{if } x < c, \end{cases}$$

where $c = \dfrac{\gamma K}{1+\gamma}$, $\gamma = \dfrac{2r}{\sigma^2}$.

x: current price.
c: trigger price.
r: interest rate.
K: exercise price.
σ: volatility.

30.9

$X_t = X_0 + \int_0^t u(s, \omega)\,ds + \int_0^t v(s, \omega)\,dB_s,$

where $P[\int_0^t v(s, \omega)^2\,ds < \infty \text{ for all } t \geq 0] = 1,$ and $P[\int_0^t |u(s, \omega)|\,ds < \infty \text{ for all } t \geq 0] = 1.$ Both u and v are adapted to the filtration $\{\mathcal{F}_t\}$, where B_t is an $\mathcal{F}_t$-*Brownian motion*.

X_t is by definition a one-dimensional *stochastic integral*.

30.10 $dX_t = u\,dt + v\,dB_t$

A differential form of (30.9).

If $dX_t = u\,dt + v\,dB_t$ and $Y_t = g(X_t)$, where g is C^2, then

30.11

$$dY_t = \left(g'(X_t)u + \tfrac{1}{2}g''(X_t)v^2\right)dt + g'(X_t)v\,dB_t$$

Itô's formula (one-dimensional).

30.12 $dt \cdot dt = dt \cdot dB_t = dB_t \cdot dt = 0, \quad dB_t \cdot dB_t = dt$

Useful relations.

30.13
$$d \ln X_t = \left(\frac{u}{X_t} - \frac{v^2}{2X_t^2} \right) dt + \frac{v}{X_t} dB_t$$

$$de^{X_t} = \left(e^{X_t} u + \frac{1}{2} e^{X_t} v^2 \right) dt + e^{X_t} v \, dB_t$$

Two special cases of (30.11).

30.14
$$\begin{pmatrix} dX_1 \\ \vdots \\ dX_n \end{pmatrix} = \begin{pmatrix} u_1 \\ \vdots \\ u_n \end{pmatrix} dt + \begin{pmatrix} v_{11} & \cdots & v_{1m} \\ \vdots & & \vdots \\ v_{n1} & \cdots & v_{nm} \end{pmatrix} \begin{pmatrix} dB_1 \\ \vdots \\ dB_m \end{pmatrix}$$

Vector version of (30.10), where $B_1, \ldots, B_m$ are m independent one-dimensional Brownian motions.

30.15
If $\mathbf{Y} = (Y_1, \ldots, Y_k) = \mathbf{g}(t, \mathbf{X})$, where $\mathbf{g} = (g_1, \ldots, g_k)$ is C^2, then for $r = 1, \ldots, k$,

$$dY_r = \frac{\partial g_r(t, \mathbf{X})}{\partial t} dt + \sum_{i=1}^{n} \frac{\partial g_r(t, \mathbf{X})}{\partial x_i} dX_i$$

$$+ \frac{1}{2} \sum_{i,j=1}^{n} \frac{\partial^2 g_r(t, \mathbf{X})}{\partial x_i \partial x_j} dX_i \, dX_j$$

where $dt \cdot dt = dt \cdot dB_i = 0$ and $dB_i \cdot dB_j = dt$ if $i = j$, 0 if $i \neq j$.

An n-dimensional version of Itô's formula.

30.16
$$J(t, x) = \max_u E^{t,x} \left[\int_t^T e^{-rs} W(s, X_s, u_s) \, ds \right],$$

where T is fixed, $u_s \in U$, U is a fixed interval, and

$$dX_t = b(t, X_t, u_t) \, dt + \sigma(t, X_t, u_t) \, dB_t.$$

A *stochastic control problem*. J is the value function, u_t is the control. $E^{t,x}$ is expectation subject to the initial condition $X_t = x$.

30.17
$$-J'_t(t, x) = \max_{u \in U} \Big[W(t, x, u)$$

$$+ J'_x(t, x) b(t, x, u) + \tfrac{1}{2} J''_{xx}(t, x) (\sigma(t, x, u))^2 \Big]$$

The *Hamilton–Jacobi–Bellman* equation. A necessary condition for optimality in (30.16).

References

For (30.1) and (30.2), see Sharpe (1964). For (30.3), see Black and Scholes (1973). For (30.5), and many other option pricing formulas, see Haug (1997), who also gives detailed references to the literature as well as computer codes for option pricing formulas. For (30.8), see Merton (1973). For stochastic calculus and stochastic control theory, see Øksendal (2003), Fleming and Rishel (1975), and Karatzas and Shreve (1991).

Chapter 31

Non-cooperative game theory

31.1 In an n-person game we assign to each player i $(i = 1, \ldots, n)$ a *strategy set* S_i and a pure strategy payoff function u_i that gives player i utility $u_i(\mathbf{s}) = u_i(s_1, \ldots, s_n)$ for each *strategy profile* $\mathbf{s} = (s_1, \ldots, s_n) \in S = S_1 \times \cdots \times S_n$.

An *n-person game* in strategic (or *normal*) form. If all the strategy sets S_i have a finite number of elements, the game is called *finite*.

31.2 A strategy profile $(s_1^*, \ldots, s_n^*)$ for an n-person game is a *pure strategy Nash equilibrium* if for all $i = 1, \ldots, n$ and all s_i in S_i,

$$u_i(s_1^*, \ldots, s_n^*) \geq u_i(s_1^*, \ldots, s_{i-1}^*, s_i, s_{i+1}^*, \ldots s_n^*)$$

Definition of a pure strategy Nash equilibrium for an n-person game.

31.3 If for all $i = 1, \ldots, n$, the strategy set S_i is a nonempty, compact, and convex subset of $\mathbb{R}^m$, and $u_i(s_1, \ldots, s_n)$ is continuous in $S = S_1 \times \cdots \times S_n$ and quasiconcave in its ith variable, then the game has a pure strategy Nash equilibrium.

Sufficient conditions for the existence of a pure strategy Nash equilibrium. (There will usually be several Nash equilibria.)

31.4 Consider a finite n-person game where S_i is player i's pure strategy set, and let $S = S_1 \times \cdots \times S_n$. Let Ω_i be a set of probability distributions over S_i. An element σ_i of Ω_i (σ_i is then a function $\sigma_i : S_i \to [0,1]$) is called a *mixed strategy* for player i, with the interpretation that if i plays σ_i, then i chooses the pure strategy s_i with probability $\sigma_i(s_i)$.

If the players choose the *mixed strategy profile* $\boldsymbol{\sigma} = (\sigma_1, \ldots, \sigma_n) \in \Omega_1 \times \cdots \times \Omega_n$, the probability that the pure strategy profile $\mathbf{s} = (s_1, \ldots, s_n)$ occurs is $\sigma_1(s_1) \cdots \sigma_n(s_n)$. The expected payoff to player i is then

$$u_i(\boldsymbol{\sigma}) = \sum_{s \in S} \sigma_1(s_1) \cdots \sigma_n(s_n) u_i(s)$$

Definition of a *mixed strategy* for an n-person game.

31.5 A mixed strategy profile $\boldsymbol{\sigma}^* = (\sigma_1^*, \ldots, \sigma_n^*)$ is a *Nash equilibrium* if for all i and every σ_i,

$$u_i(\boldsymbol{\sigma}^*) \geq u_i(\sigma_1^*, \ldots, \sigma_{i-1}^*, \sigma_i, \sigma_{i+1}^*, \ldots, \sigma_n^*)$$

Definition of a *mixed strategy Nash equilibrium* for an n-person game.

31.6 $\boldsymbol{\sigma}^*$ is a Nash equilibrium if and only if the following conditions hold for all $i = 1, \ldots, n$:

$$\sigma_i^*(s_i) > 0 \Rightarrow u_i(\boldsymbol{\sigma}^*) = u_i(s_i, \boldsymbol{\sigma}_{-i}^*) \text{ for all } s_i$$

$$\sigma_i^*(s_i') = 0 \Rightarrow u_i(\boldsymbol{\sigma}^*) \geq u_i(s_i', \boldsymbol{\sigma}_{-i}^*) \text{ for all } s_i'$$

where $\boldsymbol{\sigma}_{-i}^* = (\sigma_1^*, \ldots, \sigma_{i-1}^*, \sigma_{i+1}^*, \ldots, \sigma_n^*)$ and we consider s_i and s_i' as degenerate mixed strategies.

An equivalent definition of a (mixed strategy) Nash equilibrium.

31.7 Every finite n-person game has a mixed strategy Nash equilibrium.

An important result.

31.8 The pure strategy $s_i \in S_i$ of player i is *strictly dominated* if there exists a mixed strategy σ_i for player i such that for all feasible combinations of the other players' pure strategies, i's payoff from playing strategy s_i is strictly less than i's payoff from playing σ_i:

$$u_i(s_1, \ldots, s_{i-1}, s_i, s_{i+1}, \ldots, s_n) <$$
$$u_i(s_1, \ldots, s_{i-1}, \sigma_i, s_{i+1}, \ldots, s_n)$$

for every $(s_1, \ldots, s_{i-1}, s_{i+1}, \ldots, s_n)$ that can be constructed from the other players' strategy sets $S_1, \ldots, S_{i-1}, S_{i+1}, \ldots, S_n$.

Definition of strictly dominated strategies.

31.9 In an n-person game, the following results hold:

- If iterated elimination of strictly dominated strategies eliminates all but the strategies $(s_1^*, \ldots, s_n^*)$, then these strategies are the unique Nash equilibrium of the game.

- If the mixed strategy profile $(\sigma_1^*, \ldots, \sigma_n^*)$ is a Nash equilibrium and, for some player i, $\sigma_i^*(s_i) > 0$, then s_i survives iterated elimination of strictly dominated strategies.

Useful results. Iterated elimination of strictly dominated strategies need not result in the elimination of any strategy. (For a discussion of iterated elimination of strictly dominated strategies, see the literature.)

A two-person game where the players 1 and 2 have m and n (pure) strategies, respectively, can be represented by the two payoff matrices

31.10
$$\mathbf{A} = \begin{pmatrix} a_{11} & \cdots & a_{1n} \\ a_{21} & \cdots & a_{2n} \\ \vdots & & \vdots \\ a_{m1} & \cdots & a_{mn} \end{pmatrix}, \quad \mathbf{B} = \begin{pmatrix} b_{11} & \cdots & b_{1n} \\ b_{21} & \cdots & b_{2n} \\ \vdots & & \vdots \\ b_{m1} & \cdots & b_{mn} \end{pmatrix}$$

a_{ij} (b_{ij}) is the payoff to player 1 (2) when the players play their pure strategies i and j, respectively. If $\mathbf{A} = -\mathbf{B}$, the game is a *zero-sum* game. The game is *symmetric* if $\mathbf{A} = \mathbf{B}'$.

For the two-person game in (31.10) there exists a Nash equilibrium $(\mathbf{p}^*, \mathbf{q}^*)$ such that

31.11
- $\mathbf{p} \cdot \mathbf{Aq}^* \le \mathbf{p}^* \cdot \mathbf{Aq}^*$ for all $\mathbf{p}$ in Δ_m,

- $\mathbf{p}^* \cdot \mathbf{Bq} \le \mathbf{p}^* \cdot \mathbf{Bq}^*$ for all $\mathbf{q}$ in Δ_n.

The existence of a Nash equilibrium for a two-person game. Δ_k denotes the simplex in $\mathbb{R}^k$ consisting of all non-negative vectors whose components sum to one.

In a two-person zero-sum game $(\mathbf{A} = -\mathbf{B})$, the condition for the existence of a Nash equilibrium is equivalent to the condition that $\mathbf{p} \cdot \mathbf{Aq}$ has a *saddle point* $(\mathbf{p}^*, \mathbf{q}^*)$, i.e., for all $\mathbf{p}$ in Δ_m and all $\mathbf{q}$ in Δ_n,

31.12

$$\mathbf{p} \cdot \mathbf{Aq}^* \le \mathbf{p}^* \cdot \mathbf{Aq}^* \le \mathbf{p}^* \cdot \mathbf{Aq}$$

The *saddle point* property of the Nash equilibrium for a two-person zero-sum game.

The equilibrium payoff $v = \mathbf{p}^* \cdot \mathbf{Aq}^*$ is called the *value* of the game, and

31.13
$$v = \min_{\mathbf{q} \in \Delta_n} \max_{\mathbf{p} \in \Delta_m} \mathbf{p} \cdot \mathbf{Aq} = \max_{\mathbf{p} \in \Delta_m} \min_{\mathbf{q} \in \Delta_n} \mathbf{p} \cdot \mathbf{Aq}$$

The classical *minimax theorem* for two-person zero-sum games.

Assume that $(\mathbf{p}^*, \mathbf{q}^*)$ and $(\mathbf{p}^{**}, \mathbf{q}^{**})$ are Nash equilibria in the game (31.10). Then $(\mathbf{p}^*, \mathbf{q}^{**})$ and $(\mathbf{p}^{**}, \mathbf{q}^*)$ are also equilibrium strategy profiles.

31.14

The *rectangular* or *exchangeability* property.

Evolutionary game theory

In the symmetric two-person game of (31.10) with $\mathbf{A} = \mathbf{B}'$, a strategy $\mathbf{p}^*$ is called an *evolutionary stable strategy* if for every $\mathbf{q} \ne \mathbf{p}^*$ there exists an $\bar{\varepsilon} > 0$ such that

31.15

$$\mathbf{q} \cdot \mathbf{A}(\varepsilon \mathbf{q} + (1-\varepsilon)\mathbf{p}^*) < \mathbf{p}^* \cdot \mathbf{A}(\varepsilon \mathbf{q} + (1-\varepsilon)\mathbf{p}^*)$$

for all positive $\varepsilon < \bar{\varepsilon}$.

The value of $\bar{\varepsilon}$ may depend on $\mathbf{q}$. Biological interpretation: All animals are programmed to play $\mathbf{p}^*$. Any mutation that tries invasion with $\mathbf{q}$, has strictly lower fitness.

In the setting (31.15) the strategy $\mathbf{p}^*$ is evolutionary stable if and only if

31.16 $\qquad \mathbf{q} \cdot \mathbf{A}\mathbf{p}^* \leq \mathbf{p}^* \cdot \mathbf{A}\mathbf{p}^* \quad$ for all $\mathbf{q}$.

If $\mathbf{q} \neq \mathbf{p}^*$ and $\mathbf{q} \cdot \mathbf{A}\mathbf{p}^* = \mathbf{p}^* \cdot \mathbf{A}\mathbf{p}^*$, then

$$\mathbf{q} \cdot \mathbf{A}\mathbf{q} < \mathbf{p}^* \cdot \mathbf{A}\mathbf{q}.$$

The first condition, (the *equilibrium condition*), is equivalent to the condition for a Nash equilibrium. The second condition is called a *stability condition*.

Games of incomplete information

A *game of incomplete information* assigns to each player $i = 1, \ldots, n$ private information $\varphi_i \in \Phi_i$, a strategy set S_i of rules $s_i(\varphi_i)$ and an expected utility function

31.17 $\qquad E_\Phi[u_i(s_1(\varphi_1), \ldots, s_n(\varphi_n), \boldsymbol{\varphi})]$

(The realization of φ_i is known only to player i while the distribution $F(\Phi)$ is common knowledge, $\Phi = \Phi_1 \times \cdots \times \Phi_n$. E_Φ is the expectation over $\boldsymbol{\varphi} = (\varphi_1, \ldots, \varphi_n)$.)

Informally, a game of incomplete information is one where some players do not know the payoffs to the others.)

A strategy profile s^* is a *dominant strategy equilibrium* if for all $i = 1, \ldots, n$,

31.18 $\qquad u_i(s_1(\varphi_1), \ldots, s_i^*(\varphi_i) \ldots, s_n(\varphi_n), \boldsymbol{\varphi})$
$\qquad \geq u_i(s_1(\varphi_1), \ldots, s_i(\varphi_i), \ldots, s_n(\varphi_n), \boldsymbol{\varphi})$

for all $\boldsymbol{\varphi}$ in Φ and all $\mathbf{s} = (s_1, \ldots, s_n)$ in $S = S_1 \times \cdots \times S_n$.

Two common solution concepts are dominant strategy equilibrium and Bayesian Nash equlibrium.

A strategy profile $\mathbf{s}^*$ is a *pure strategy Bayesian Nash equilibrium* if for all $i = 1, \ldots, n$,

31.19 $\qquad E_\Phi[u_1(s_1^*(\varphi_1), \ldots, s_i^*(\varphi_i), \ldots, s_n^*(\varphi_n), \boldsymbol{\varphi})]$
$\qquad \geq E_\Phi[u_1(s_1^*(\varphi_1), \ldots, s_i(\varphi_i), \ldots, s_n^*(\varphi_n), \boldsymbol{\varphi})]$
for all s_i in S_i.

Pure strategy Bayesian Nash equilibrium.

References

Friedman (1986) is a standard reference. See also Gibbons (1992) (the simplest treatment), Kreps (1990), and Fudenberg and Tirole (1991). For evolutionary game theory, see Weibull (1995). For games of incomplete information, see Mas-Colell, Whinston, and Green (1995).

Chapter 32

Combinatorics

32.1 The number of ways that n objects can be arranged in order is $n! = 1 \cdot 2 \cdot 3 \cdots (n-1) \cdot n$.

5 persons A, B, C, D, and E can be lined up in $5! = 120$ different ways.

32.2 The number of possible ordered subsets of k objects from a set of n objects, is

$$\frac{n!}{(n-k)!} = n(n-1)\cdots(n-k+1)$$

If a lottery has n tickets and k distinct prizes, there are $\frac{n!}{(n-k)!}$ possible lists of prizes.

32.3 Given a collection $S_1, S_2, \ldots, S_n$ of disjoint sets containing $k_1, k_2, \ldots, k_n$ objects, respectively, there are $k_1 k_2 \cdots k_n$ ways of selecting one object from each set.

If a restaurant has 3 different appetizers, 5 main courses, and 4 desserts, then the total number of possible dinners is $3 \cdot 5 \cdot 4 = 60$.

32.4 A set of n elements has $\binom{n}{k} = \frac{n!}{k!(n-k)!}$ different subsets of k elements.

In a card game you receive 5 cards out of 52. The number of different hands is $\binom{52}{5} = \frac{52!}{5!47!} = 2\,598\,960$.

32.5 The number of ways of arranging n objects of k different types where there are n_1 objects of type 1, n_2 objects of type 2, $\ldots$, and n_k objects of type k is $\frac{n!}{n_1! \cdot n_2! \cdots n_k!}$.

There are $\frac{12!}{5! \cdot 4! \cdot 3!} = 27\,720$ different ways that 12 persons can be allocated to three taxis with 5 in the first, 4 in the second, and 3 in the third.

Let $|X|$ denote the number of elements of a set X. Then

32.6
- $|A \cup B| = |A| + |B| - |A \cap B|$
- $|A \cup B \cup C| = |A| + |B| + |C| - |A \cap B| -$
  $|A \cap C| - |B \cap B| + |A \cap B \cap C|$

The *inclusion–exclusion principle*, special cases.

32.7
$$|A_1 \cup A_2 \cup \cdots \cup A_n| = |A_1| + |A_2| + \cdots + |A_n|$$
$$- |A_1 \cap A_2| - |A_1 \cap A_3| - \cdots - |A_{n-1} \cap A_n|$$
$$+ \cdots + (-1)^{n+1}|A_1 \cap A_2 \cap \cdots \cap A_n|$$
$$= \sum (-1)^{r+1}|A_{j_1} \cap A_{j_2} \cap \cdots \cap A_{j_r}|.$$

The sum is taken over all nonempty subsets $\{j_1, j_2, \ldots, j_r\}$ of the index set $\{1, 2, \ldots, n\}$.

The *inclusion–exclusion principle*.

32.8
If more than k objects are distributed among k boxes (pigeonholes), then some box must contain at least 2 objects. More generally, if at least $nk + 1$ objects are distributed among k boxes (pigeonholes), then some box must contain at least $n + 1$ objects.

Dirichlet's pigeonhole principle. (If $16 = 5 \cdot 3 + 1$ socks are distributed among 3 drawers, then at least one drawer must contain at least 6 socks.)

References

See e.g. Anderson (1987) or Charalambides (2002).

Chapter 33

Probability and statistics

The probability $P(A)$ of an event $A \subset \Omega$ satisfies the following axioms:

33.1

(a) $0 \leq P(A) \leq 1$

(b) $P(\Omega) = 1$

(c) If $A_i \cap A_j = \varnothing$ for $i \neq j$, then

$$P(\bigcup_{i=1}^{\infty} A_i) = \sum_{i=1}^{\infty} P(A_i)$$

Axioms for probability. Ω is the sample space consisting of all possible outcomes. An event is a subset of Ω.

  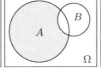

$A \cup B$	$A \cap B$	$A \setminus B$	A^c	$A \triangle B$
A or B occurs	Both A and B occur	A occurs, but B does not	A does not occur	A or B occurs, but not both

33.2

- $P(A^c) = 1 - P(A)$
- $P(A \cup B) = P(A) + P(B) - P(A \cap B)$
- $P(A \cup B \cup C) = P(A) + P(B) + P(C)$
  $\qquad - P(A \cap B) - P(A \cap C) - P(B \cap C)$
  $\qquad + P(A \cap B \cap C)$
- $P(A \setminus B) = P(A) - P(A \cap B)$
- $P(A \triangle B) = P(A) + P(B) - 2P(A \cap B)$

Rules for calculating probabilities.

33.3

$P(A \mid B) = \dfrac{P(A \cap B)}{P(B)}$ is the *conditional probability* that event A will occur given that B has occurred.

Definition of *conditional probability*, $P(B) > 0$.

33.4

A and B are *(stochastically) independent* if

$\qquad P(A \cap B) = P(A)P(B)$

If $P(B) > 0$, this is equivalent to

$\qquad P(A \mid B) = P(A)$

Definition of (stochastic) independence.

33.5
$$P(A_1 \cap A_2 \cap \ldots \cap A_n) =$$
$$P(A_1)P(A_2 \mid A_1) \cdots P(A_n \mid A_1 \cap A_2 \cap \cdots \cap A_{n-1})$$

General multiplication rule for probabilities.

33.6
$$P(A \mid B) = \frac{P(B \mid A) \cdot P(A)}{P(B)}$$
$$= \frac{P(B \mid A)P(A)}{P(B \mid A)P(A) + P(B \mid A^c)P(A^c)}$$

Bayes's rule.
($P(B) \neq 0$.)

33.7
$$P(A_i \mid B) = \frac{P(B \mid A_i) \cdot P(A_i)}{\sum\limits_{j=1}^{n} P(B \mid A_j) \cdot P(A_j)}$$

Generalized *Bayes's rule.* $A_1, \ldots, A_n$ are disjoint, $\sum_{i=1}^{n} P(A_i) = P(\Omega) = 1$, where $\Omega = \bigcup_{i=1}^{n} A_i$ is the sample space. B is an arbitrary event.

One-dimensional random variables

33.8
- $P(X \in A) = \sum\limits_{x \in A} f(x)$
- $P(X \in A) = \int\limits_{A} f(x)\, dx$

f is the discrete/continuous probability density function for the random (or stochastic) variable X.

33.9
- $F(x) = P(X \leq x) = \sum\limits_{t \leq x} f(t)$
- $F(x) = P(X \leq x) = \int\limits_{-\infty}^{x} f(t)\, dt$

F is the cumulative discrete/continuous distribution function. In the continuous case, $P(X = x) = 0$.

33.10
- $E[X] = \sum\limits_{x} x f(x)$
- $E[X] = \int\limits_{-\infty}^{\infty} x f(x)\, dx$

Expectation of a random variable X with discrete/continuous probability density function f. $\mu = E[X]$ is called the *mean.*

33.11
- $E[g(X)] = \sum\limits_{x} g(x) f(x)$
- $E[g(X)] = \int\limits_{-\infty}^{\infty} g(x) f(x)\, dx$

Expectation of a function g of a random variable X with discrete/continuous probability density function f.

33.12
$$\mathrm{var}[X] = E[(X - E[X])^2]$$

The *variance* of a random variable is, by definition, the expected value of its squared deviation from the mean.

33.13 $\mathrm{var}[X] = E[X^2] - (E[X])^2$

Another expression for the variance.

33.14 $\sigma = \sqrt{\mathrm{var}[X]}$

The *standard deviation* of X.

33.15 $\mathrm{var}[aX + b] = a^2\,\mathrm{var}[X]$

a and b are real numbers.

33.16 $\mu_k = E[(X - \mu)^k]$

The kth *central moment* about the mean, $\mu = E[X]$.

33.17 $\eta_3 = \dfrac{\mu_3}{\sigma^3}, \quad \eta_4 = \dfrac{\mu_4}{\sigma^4} - 3$

The *coefficient of skewness*, η_3, and the *coefficient of kurtosis*, η_4. σ is the standard deviation. For the normal distribution, $\eta_3 = \eta_4 = 0$.

33.18
- $P(|X| \geq \lambda) \leq E[X^2]/\lambda^2$
- $P(|X - \mu| \geq \lambda) \leq \sigma^2/\lambda^2, \quad \lambda > 0$
- $P(|X - \mu| \geq k\sigma) \leq 1/k^2, \quad k > 0$

Different versions of *Chebyshev's inequality.* σ is the standard deviation of X, $\mu = E[X]$ is the mean.

33.19 If f is convex on the interval I and X is a random variable with finite expectation, then

$$f(E[X]) \leq E[f(X)]$$

If f is strictly convex, the inequality is strict unless X is a constant with probability 1.

Special case of Jensen's inequality.

33.20
- $M(t) = E[e^{tX}] = \sum_x e^{tx} f(x)$
- $M(t) = E[e^{tX}] = \int\limits_{-\infty}^{\infty} e^{tx} f(x)\,dx$

Moment generating functions. $M(t)$ does not always exist, but if it does, then
$$M(t) = \sum_{k=0}^{\infty} \frac{E[X^k]}{k!} t^k.$$

33.21 If the moment generating function $M(t)$ defined in (33.20) exists in an open neighborhood of 0, then $M(t)$ uniquely determines the probability distribution function.

An important result.

33.22
- $C(t) = E[e^{itX}] = \sum_x e^{itx} f(x)$
- $C(t) = E[e^{itX}] = \int\limits_{-\infty}^{\infty} e^{itx} f(x)\,dx$

Characteristic functions. $C(t)$ always exists, and if $E[X^k]$ exists for all k, then
$$C(t) = \sum_{k=0}^{\infty} \frac{i^k E[X^k]}{k!} t^k.$$

| 33.23 | The characteristic function $C(t)$ defined in (33.22) uniquely determines the probability distribution function $f(x)$. | An important result. |

Two-dimensional random variables

| 33.24 | • $P((X,Y) \in A) = \sum\limits_{(x,y) \in A} f(x,y)$

 • $P((X,Y) \in A) = \iint\limits_{A} f(x,y)\,dx\,dy$ | $f(x,y)$ is the two-dimensional discrete/continuous *simultaneous density function* for the random variables X and Y. |

| 33.25 | $F(x,y) = P(X \le x, Y \le y) =$

 • $\sum\limits_{u \le x} \sum\limits_{v \le y} f(u,v)$ (discrete case)

 • $\int\limits_{-\infty}^{x} \int\limits_{-\infty}^{y} f(u,v)\,du\,dv$ (continuous case) | F is the simultaneous cumulative discrete/continuous distribution function. |

| 33.26 | • $E[g(X,Y)] = \sum\limits_{x} \sum\limits_{y} g(x,y) f(x,y)$

 • $E[g(X,Y)] = \int\limits_{-\infty}^{\infty} \int\limits_{-\infty}^{\infty} g(x,y) f(x,y)\,dx\,dy$ | The expectation of $g(X,Y)$, where X and Y have the simultaneous discrete/continuous density function f. |

| 33.27 | $\operatorname{cov}[X,Y] = E\big[(X - E[X])(Y - E[Y])\big]$ | Definition of *covariance*. |

| 33.28 | $\operatorname{cov}[X,Y] = E[XY] - E[X]\,E[Y]$ | A useful fact. |

| 33.29 | If $\operatorname{cov}[X,Y] = 0$, X and Y are *uncorrelated*. | A definition. |

| 33.30 | $E[XY] = E[X]\,E[Y]$ if X and Y are uncorrelated. | Follows from (33.28) and (33.29). |

| 33.31 | $(E[XY])^2 \le E[X^2]\,E[Y^2]$ | Cauchy–Schwarz's inequality. |

| 33.32 | If X and Y are stochastically independent, then $\operatorname{cov}[X,Y] = 0$. | The converse is not true. |

| 33.33 | $\operatorname{var}[X \pm Y] = \operatorname{var}[X] + \operatorname{var}[Y] \pm 2\operatorname{cov}[X,Y]$ | The variance of a sum/difference of two random variables. |

| 33.34 | $E[a_1 X_1 + \cdots + a_n X_n + b] =$
 $\qquad a_1 E[X_1] + \cdots + a_n E[X_n] + b$ | $X_1, \ldots, X_n$ are random variables and $a_1, \ldots, a_n$, b are real numbers. |

33.35	$$\mathrm{var}\Big[\sum_{i=1}^{n} a_i X_i\Big] = \sum_{i=1}^{n}\sum_{j=1}^{n} a_i a_j \,\mathrm{cov}[X_i, X_j]$$ $$= \sum_{i=1}^{n} a_i^2 \,\mathrm{var}[X_i] + 2\sum_{i=1}^{n-1}\sum_{j=i+1}^{n} a_i a_j \,\mathrm{cov}[X_i, X_j]$$	The variance of a linear combination of random variables.

33.36	$$\mathrm{var}\Big[\sum_{i=1}^{n} a_i X_i\Big] = \sum_{i=1}^{n} a_i^2 \,\mathrm{var}[X_i]$$	Formula (33.35) when $X_1, \ldots, X_n$ are pairwise uncorrelated.

33.37	$$\mathrm{corr}[X,Y] = \frac{\mathrm{cov}[X,Y]}{\sqrt{\mathrm{var}[X]\,\mathrm{var}[Y]}} \in [-1,1]$$	Definition of the *correlation coefficient* as a normalized covariance.

33.38	If $f(x,y)$ is a simultaneous density distribution function for X and Y, then • $f_X(x) = \sum_y f(x,y), \quad f_Y(y) = \sum_x f(x,y)$ • $f_X(x) = \int_{-\infty}^{\infty} f(x,y)\,dy, \quad f_Y(y) = \int_{-\infty}^{\infty} f(x,y)\,dx$ are the *marginal densities* of X and Y, respectively.	Definitions of *marginal densities* for discrete and continuous distributions.

33.39	$$f(x\,|\,y) = \frac{f(x,y)}{f_Y(y)}, \quad f(y\,|\,x) = \frac{f(x,y)}{f_X(x)}$$	Definitions of *conditional densities*.

33.40	The random variables X and Y are *stochastically independent* if $f(x,y) = f_X(x)f_Y(y)$. If $f_Y(y) > 0$, this is equivalent to $f(x\,|\,y) = f_X(x)$.	Stochastic independence.

33.41	• $E[X\,|\,y] = \sum_x x f(x\,|\,y)$ • $E[X\,|\,y] = \int_{-\infty}^{\infty} x f(x\,|\,y)\,dx$ • $\mathrm{var}[X\,|\,y] = \sum_x \big(x - E[X\,|\,y]\big)^2 f(x\,|\,y)$ • $\mathrm{var}[X\,|\,y] = \int_{-\infty}^{\infty} (x - E[X\,|\,y])^2 f(x\,|\,y)\,dx$	Definitions of *conditional expectation* and *conditional variance* for discrete and continuous distributions. Note that $E[X\,|\,y]$ denotes $E[X\,|\,Y = y]$, and $\mathrm{var}[X\,|\,y]$ denotes $\mathrm{var}[X\,|\,Y = y]$.

33.42	$$E[Y] = E_X[E[Y\,|\,X]]$$	Law of *iterated expectations*. E_X denotes expectation w.r.t. X.

33.43 $\quad E[XY] = E_X[XE[Y\,|\,X]] = E[X\mu_{Y|X}].$

> The expectation of XY is equal to the expected product of X and the conditional expectation of Y given X.

33.44
$$\sigma_Y^2 = \mathrm{var}[Y] = E_X[\mathrm{var}[Y\,|\,X]] + \mathrm{var}_X[E[Y\,|\,X]]$$
$$= E[\sigma_{Y|X}^2] + \mathrm{var}[\mu_{Y|X}]$$

> The variance of Y is equal to the expectation of its conditional variances plus the variance of its conditional expectations.

Let $f(x,y)$ be the density function for a pair (X,Y) of stochastic variables. Suppose that
$$U = \varphi_1(X,Y), \quad V = \varphi_2(X,Y)$$
is a one-to-one C^1 transformation of (X,Y), with the inverse transformation given by
$$X = \psi_1(U,V), \quad Y = \psi_2(U,V)$$

33.45 Then the density function $g(u,v)$ for the pair (U,V) is given by
$$g(u,v) = f(\psi_1(u,v), \psi_2(u,v)) \cdot |J(u,v)|$$
provided
$$J(u,v) = \begin{vmatrix} \dfrac{\partial \psi_1(u,v)}{\partial u} & \dfrac{\partial \psi_1(u,v)}{\partial v} \\[2mm] \dfrac{\partial \psi_2(u,v)}{\partial u} & \dfrac{\partial \psi_2(u,v)}{\partial v} \end{vmatrix} \neq 0$$

> How to find the density function of a transformation of stochastic variables. (The formula generalizes in a straightforward manner to the case with an arbitrary number of stochastic variables. The required regularity conditions are not fully spelled out. See the references.) $J(u,v)$ is the Jacobian determinant.

Statistical inference

33.46 If $E[\hat{\theta}] = \theta$ for all θ in Θ, then $\hat{\theta}$ is called an *unbiased estimator* of θ.

> Definition of an unbiased estimator. Θ is the parameter space.

If $\hat{\theta}$ is not unbiased,
33.47 $\quad b = E[\hat{\theta}] - \theta$
is called the *bias* of $\hat{\theta}$.

> Definition of bias.

33.48 $\quad \mathrm{MSE}(\hat{\theta}) = E[\hat{\theta} - \theta]^2 = \mathrm{var}[\hat{\theta}] + b^2$

> Definition of *mean square error*, MSE.

33.49 $\mathrm{plim}\,\hat{\theta}_T = \theta$ means that for every $\varepsilon > 0$
$$\lim_{T \to \infty} P(|\hat{\theta}_T - \theta| < \varepsilon) = 1$$

> Definition of a probability limit. The estimator $\hat{\theta}_T$ is a function of T observations.

33.50	If θ_T has mean μ_T and variance σ_T^2 such that the ordinary limits of μ_T and σ_T^2 are θ and 0 respectively, then θ_T *converges in mean square* to θ, and $\operatorname{plim}\hat{\theta}_T = \theta$.	Convergence in quadratic mean (mean square convergence).
33.51	If g is continuous, then $$\operatorname{plim} g(\theta_T) = g(\operatorname{plim}\theta_T)$$	Slutsky's theorem.
33.52	If θ_T and ω_T are random variables with probability limits $\operatorname{plim}\theta_T = \theta$ and $\operatorname{plim}\omega_T = \omega$, then • $\operatorname{plim}(\theta_T + \omega_T) = \theta + \omega$ • $\operatorname{plim}(\theta_T\omega_T) = \theta\omega$ • $\operatorname{plim}(\theta_T/\omega_T) = \theta/\omega$	Rules for probability limits.
33.53	θ_T *converges in distribution* to a random variable θ with cumulative distribution function F if $\lim_{T\to\infty}\|F_T(\theta) - F(\theta)\| = 0$ at all continuity points of $F(\theta)$. This is written: $$\theta_T \xrightarrow{d} \theta$$	Limiting distribution.
33.54	If $\theta_T \xrightarrow{d} \theta$ and $\operatorname{plim}(\omega_T) = \omega$, then • $\theta_T\omega_T \xrightarrow{d} \theta\omega$ • If ω_T has a limiting distribution and the limit $\operatorname{plim}(\theta_T - \omega_T) = 0$, then θ_T has the same limiting distribution as ω_T.	Rules for limiting distributions.
33.55	$\hat{\theta}$ is a *consistent* estimator of θ if $$\operatorname{plim}\hat{\theta}_T = \theta \text{ for every } \theta \in \Theta.$$	Definition of consistency.
33.56	$\hat{\theta}$ is *asymptotically unbiased* if $$\lim_{T\to\infty} E[\hat{\theta}_T] = \theta \text{ for every } \theta \in \Theta.$$	Definition of an asymptotically unbiased estimator.
33.57	H_0 Null hypothesis (e.g. $\theta \le 0$). H_1 Alternate hypothesis (e.g. $\theta > 0$). T Test statistic. C Critical region. θ An unknown parameter.	Definitions for *statistical testing*.
33.58	A test: Reject H_0 in favor of H_1 if $T \in C$.	A test.
33.59	The power function of a test is $$\pi(\theta) = P(\text{reject } H_0 \mid \theta), \theta \in \Theta.$$	Definition of the *power* of a test.

33.60 To reject H_0 when H_0 is true is called a type I error.

Not to reject H_0 when H_1 is true is called a type II error.

Type I and II errors.

33.61 α-level of significance: The least α such that P(type I error) $\leq \alpha$ for all θ satisfying H_0.

The α-level of significance of a test.

33.62 *Significance probability* (or *P-value*) is the least level of significance that leads to rejection of H_0, given the data and the test.

An important concept.

Asymptotic results

33.63 Let $\{X_i\}$ be a sequence of independent and identically distributed random variables, with finite mean $E[X_i] = \mu$. Let $S_n = X_1 + \cdots + X_n$. Then:

(1) For every $\varepsilon > 0$,

$$P\left\{\left|\frac{S_n}{n} - \mu\right| < \varepsilon\right\} \to 1 \text{ as } n \to \infty.$$

(2) With probability 1, $\dfrac{S_n}{n} \to \mu$ as $n \to \infty$.

(1) is the *weak law of large numbers*. S_n/n is a consistent estimator for μ. (2) is the *strong law of large numbers*.

33.64 Let $\{X_i\}$ be a sequence of independent and identically distributed random variables with finite mean $E[X_i] = \mu$ and finite variance $\mathrm{var}[X_i] = \sigma^2$. Let $S_n = X_1 + \cdots + X_n$. Then the distribution of $\dfrac{S_n - n\mu}{\sigma\sqrt{n}}$ tends to the standard normal distribution as $n \to \infty$, i.e.

$$P\left\{\frac{S_n - n\mu}{\sigma\sqrt{n}} \leq a\right\} \to \frac{1}{\sqrt{2\pi}} \int_{-\infty}^{a} e^{-x^2/2} \, dx$$

as $n \to \infty$.

The *central limit theorem*.

References

See Johnson and Bhattacharyya (1996), Larsen and Marx (1986), Griffiths, Carter, and Judge (1993), Rice (1995), and Hogg and Craig (1995).

Chapter 34

Probability distributions

34.1

$$f(x) = \begin{cases} \dfrac{x^{p-1}(1-x)^{q-1}}{B(p,q)}, & x \in (0,1), \\ 0 & \text{otherwise,} \end{cases}$$

$p > 0$, $q > 0$.

Mean: $E[X] = \dfrac{p}{p+q}$.

Variance: $\text{var}[X] = \dfrac{pq}{(p+q)^2(p+q+1)}$.

kth moment: $E[X^k] = \dfrac{B(p+k,q)}{B(p,q)}$.

Beta distribution. B is the beta function defined in (9.61).

34.2

$$f(x) = \binom{n}{x} p^x (1-p)^{n-x},$$

$x = 0, 1, \ldots, n$; $n = 1, 2, \ldots$; $p \in (0,1)$.

Mean: $E[X] = np$.
Variance: $\text{var}[X] = np(1-p)$.
Moment generating function: $[pe^t + (1-p)]^n$.
Characteristic function: $[pe^{it} + (1-p)]^n$.

Binomial distribution. $f(x)$ is the probability for an event to occur exactly x times in n independent observations, when the probability of the event is p at each observation. For $\binom{n}{x}$, see (8.30).

34.3

$$f(x,y) = \dfrac{e^{-Q}}{2\pi\sigma\tau\sqrt{1-\rho^2}}, \quad \text{where}$$

$$Q = \dfrac{\left(\frac{x-\mu}{\sigma}\right)^2 - 2\rho\frac{(x-\mu)(y-\eta)}{\sigma\tau} + \left(\frac{y-\eta}{\tau}\right)^2}{2(1-\rho^2)},$$

$x, y, \mu, \eta \in (-\infty, \infty)$, $\sigma > 0$, $\tau > 0$, $|\rho| < 1$.

Mean: $E[X] = \mu$, $E[Y] = \eta$.
Variance: $\text{var}[X] = \sigma^2$, $\text{var}[Y] = \tau^2$.
Covariance: $\text{cov}[X,Y] = \rho\sigma\tau$.

Binormal distribution. (For moment generating and characteristic functions, see the more general multivariate normal distribution in (34.15).)

34.4

$$f(x) = \begin{cases} \dfrac{x^{\frac{1}{2}\nu-1}e^{-\frac{1}{2}x}}{2^{\frac{1}{2}\nu}\Gamma(\frac{1}{2}\nu)}, & x > 0 \\ 0, & x \le 0 \end{cases} \quad ; \quad \nu = 1, 2, \ldots$$

Mean: $E[X] = \nu$.

Variance: $\mathrm{var}[X] = 2\nu$.

Moment generating function: $(1 - 2t)^{-\frac{1}{2}\nu}$, $t < \frac{1}{2}$.

Characteristic function: $(1 - 2it)^{-\frac{1}{2}\nu}$.

Chi-square distribution with ν degrees of freedom. Γ is the gamma function defined in (9.53).

34.5

$$f(x) = \begin{cases} \lambda e^{-\lambda x}, & x > 0 \\ 0, & x \le 0 \end{cases} \quad ; \quad \lambda > 0.$$

Mean: $E[X] = 1/\lambda$.

Variance: $\mathrm{var}[X] = 1/\lambda^2$.

Moment generating function: $\lambda/(\lambda - t)$, $t < \lambda$.

Characteristic function: $\lambda/(\lambda - it)$.

Exponential distribution.

34.6

$$f(x) = \frac{1}{\beta}e^{-z}e^{-e^{-z}}, \quad z = \frac{x - \alpha}{\beta}, \quad x \in \mathbb{R}, \quad \beta > 0$$

Mean: $E[X] = \alpha - \beta\Gamma'(1)$.

Variance: $\mathrm{var}[X] = \beta^2\pi^2/6$.

Moment gen. function: $e^{\alpha t}\Gamma(1 - \beta t)$, $t < 1/\beta$.

Characteristic function: $e^{i\alpha t}\Gamma(1 - i\beta t)$.

Extreme value (Gumbel) distribution. $\Gamma'(1)$ is the derivative of the gamma function at 1. (See (9.53).) $\Gamma'(1) = -\gamma$, where $\gamma \approx 0.5772$ is Euler's constant, see (8.48).

34.7

$$f(x) = \begin{cases} \dfrac{\nu_1^{\frac{1}{2}\nu_1}\nu_2^{\frac{1}{2}\nu_2}x^{\frac{1}{2}\nu_1-1}}{B(\frac{1}{2}\nu_1, \frac{1}{2}\nu_2)(\nu_2 + \nu_1 x)^{\frac{1}{2}(\nu_1+\nu_2)}}, & x > 0 \\ 0, & x \le 0 \end{cases}$$

$\nu_1, \nu_2 = 1, 2, \ldots$

Mean: $E[X] = \nu_2/(\nu_2 - 2)$ for $\nu_2 > 2$

(does not exist for $\nu_2 = 1, 2$).

Variance: $\mathrm{var}[X] = \dfrac{2\nu_2^2(\nu_1 + \nu_2 - 2)}{\nu_1(\nu_2 - 2)^2(\nu_2 - 4)}$ for $\nu_2 > 4$

(does not exist for $\nu_2 \le 4$).

kth moment:

$$E[X^k] = \frac{\Gamma(\frac{1}{2}\nu_1 + k)\Gamma(\frac{1}{2}\nu_2 - k)}{\Gamma(\frac{1}{2}\nu_1)\Gamma(\frac{1}{2}\nu_2)}\left(\frac{\nu_2}{\nu_1}\right)^k, \quad 2k < \nu_2$$

F-distribution. B is the beta function defined in (9.61). ν_1, ν_2 are the degrees of freedom for the numerator and denominator, respectively.

34.8

$$f(x) = \begin{cases} \dfrac{\lambda^n x^{n-1}e^{-\lambda x}}{\Gamma(n)}, & x > 0 \\ 0, & x \le 0 \end{cases} \quad ; \quad n, \lambda > 0.$$

Mean: $E[X] = n/\lambda$.

Variance: $\mathrm{var}[X] = n/\lambda^2$.

Moment generating function: $[\lambda/(\lambda - t)]^n$, $t < \lambda$.

Characteristic function: $[\lambda/(\lambda - it)]^n$.

Gamma distribution. Γ is the gamma function defined in (9.53). For $n = 1$ this is the exponential distribution.

$$f(x) = p(1-p)^x; \quad x = 0, 1, 2, \ldots, \quad p \in (0,1).$$

34.9

Mean: $E[X] = (1-p)/p$.
Variance: $\text{var}[X] = (1-p)/p^2$.
Moment generating function:
$$p/[1 - (1-p)e^t], \quad t < -\ln(1-p).$$
Characteristic function: $p/[1 - (1-p)e^{it}]$.

Geometric distribution.

34.10

$$f(x) = \frac{\binom{M}{x}\binom{N-M}{n-x}}{\binom{N}{n}},$$

$$x = 0, 1, \ldots, n; \quad n = 1, 2, \ldots, N.$$

Mean: $E[X] = nM/N$.
Variance: $\text{var}[X] = np(1-p)(N-n)/(N-1)$,
  where $p = M/N$.

Hypergeometric distribution. Given a collection of N objects, where M objects have a certain characteristic and $N-M$ do not have it. Pick n objects at random from the collection. $f(x)$ is then the probability that x objects have the characteristic and $n-x$ do not have it.

34.11

$$f(x) = \frac{1}{2\beta}\, e^{-|x-\alpha|/\beta}; \quad x \in \mathbb{R}, \quad \beta > 0$$

Mean: $E[X] = \alpha$.
Variance: $\text{var}[X] = 2\beta^2$.

Moment gen. function: $\dfrac{e^{\alpha t}}{1 - \beta^2 t^2}, \quad |t| < 1/\beta.$

Characteristic function: $\dfrac{e^{i\alpha t}}{1 + \beta^2 t^2}.$

Laplace distribution.

34.12

$$f(x) = \frac{e^{-z}}{\beta(1 + e^{-z})^2}, \quad z = \frac{x-\alpha}{\beta}, \quad x \in \mathbb{R}, \quad \beta > 0$$

Mean: $E[X] = \alpha$.
Variance: $\text{var}[X] = \pi^2\beta^2/3$.
Moment generating function:
$$e^{\alpha t}\Gamma(1 - \beta t)\Gamma(1 + \beta t) = \pi\beta t e^{\alpha t}/\sin(\pi\beta t).$$
Characteristic function: $i\pi\beta t e^{i\alpha t}/\sin(i\pi\beta t)$.

Logistic distribution.

34.13

$$f(x) = \begin{cases} \dfrac{e^{-(\ln x - \mu)^2/2\sigma^2}}{\sigma x \sqrt{2\pi}}, & x > 0 \\ 0, & x \le 0 \end{cases}; \quad \sigma > 0$$

Mean: $E[X] = e^{\mu + \frac{1}{2}\sigma^2}$.
Variance: $\text{var}[X] = e^{2\mu}(e^{2\sigma^2} - e^{\sigma^2})$.
kth moment: $E[X^k] = e^{k\mu + \frac{1}{2}k^2\sigma^2}$.

Lognormal distribution.

$$f(\mathbf{x}) = \frac{n!}{x_1! \cdots x_k!} p_1^{x_1} \cdots p_k^{x_k}$$

$x_1 + \cdots + x_k = n, \quad p_1 + \cdots + p_k = 1,$

$x_j \in \{0, 1, \ldots, n\}, \quad p_j \in (0, 1), \quad j = 1, \ldots, k.$

Mean of X_j: $E[X_j] = np_j$.

34.14   Variance of X_j: $\mathrm{var}[X_j] = np_j(1 - p_j)$.

Covariance: $\mathrm{cov}[X_j, X_r] = -np_j p_r,$

$$j, r = 1, \ldots, n, \ j \neq r.$$

Moment generating function: $\left[\sum_{j=1}^{k} p_j e^{t_j}\right]^n$.

Characteristic function: $\left[\sum_{j=1}^{k} p_j e^{it_j}\right]^n$.

Multinomial distribution. $f(\mathbf{x})$ is the probability for k events $A_1, \ldots, A_k$ to occur exactly $x_1, \ldots, x_k$ times in n independent observations, when the probabilities of the events are $p_1, \ldots, p_k$.

$$f(\mathbf{x}) = \frac{1}{(2\pi)^{k/2} \sqrt{|\Sigma|}} e^{-\frac{1}{2}(\mathbf{x}-\boldsymbol{\mu})' \Sigma^{-1}(\mathbf{x}-\boldsymbol{\mu})}$$

$\Sigma = (\sigma_{ij})$ is symmetric and positive definite,

$\mathbf{x} = (x_1, \ldots, x_k)', \ \boldsymbol{\mu} = (\mu_1, \ldots, \mu_k)'.$

34.15

Mean: $E[X_i] = \mu_i$.

Variance: $\mathrm{var}[X_i] = \sigma_{ii}$.

Covariance: $\mathrm{cov}[X_i, X_j] = \sigma_{ij}$.

Moment generating function: $e^{\boldsymbol{\mu}' \mathbf{t} + \frac{1}{2} \mathbf{t}' \Sigma \mathbf{t}}$.

Characteristic function: $e^{-\frac{1}{2} \mathbf{t}' \Sigma \mathbf{t}} e^{it'\boldsymbol{\mu}}$.

Multivariate normal distribution. $|\Sigma|$ denotes the determinant of the variance-covariance matrix Σ. $\mathbf{x} = (x_1, \ldots, x_k)'$, $\boldsymbol{\mu} = (\mu_1, \ldots, \mu_k)'$.

$$f(x) = \binom{x-1}{r-1} p^r (1-p)^{x-r},$$

$x = r, r+1, \ldots; \quad r = 1, 2, \ldots; \quad p \in (0, 1).$

34.16

Mean: $E[X] = r/p$.

Variance: $\mathrm{var}[X] = r(1-p)/p^2$.

Moment generating function: $p^r (1 - (1-p)e^t)^{-r}$.

Characteristic function: $p^r (1 - (1-p)e^{it})^{-r}$.

Negative binomial distribution.

$$f(x) = \frac{1}{\sigma\sqrt{2\pi}} e^{-(x-\mu)^2/2\sigma^2}, \quad x \in \mathbb{R}, \ \sigma > 0.$$

34.17

Mean: $E[X] = \mu$.

Variance: $\mathrm{var}[X] = \sigma^2$.

Moment generating function: $e^{\mu t + \frac{1}{2}\sigma^2 t^2}$.

Characteristic function: $e^{i\mu t - \frac{1}{2}\sigma^2 t^2}$.

Normal distribution. If $\mu = 0$ and $\sigma = 1$, this is the *standard normal distribution*.

$$f(x) = \begin{cases} \dfrac{ca^c}{x^{c+1}}, & x > a \\ 0, & x \leq a \end{cases}; \quad a > 0, \ c > 0.$$

34.18   Mean: $E[X] = ac/(c-1), \quad c > 1.$

Variance: $\mathrm{var}[X] = a^2 c/(c-1)^2(c-2), \quad c > 2.$

kth moment: $E[X^k] = a^k c/c - k, \quad c > k.$

Pareto distribution.

$$f(x) = e^{-\lambda}\frac{\lambda^x}{x!}; \quad x = 0, 1, 2, \ldots, \quad \lambda > 0.$$

34.19   Mean: $E[X] = \lambda$.   *Poisson distribution.*
Variance: $\text{var}[X] = \lambda$.
Moment generating function: $e^{\lambda(e^t - 1)}$.
Characteristic function: $e^{\lambda(e^{it} - 1)}$.

$$f(x) = \frac{\Gamma(\frac{1}{2}(\nu + 1))}{\sqrt{\nu\pi}\,\Gamma(\frac{1}{2}\nu)}\left(1 + \frac{x^2}{\nu}\right)^{-\frac{1}{2}(\nu + 1)},$$

$x \in \mathbb{R}, \quad \nu = 1, 2, \ldots$

Mean: $E[X] = 0$ for $\nu > 1$
(does not exist for $\nu = 1$).   *Student's t-distribution*

34.20   Variance: $\text{var}[X] = \dfrac{\nu}{\nu - 2}$ for $\nu > 2$ *with ν degrees of free-*
dom.
(does not exist for $\nu = 1, 2$).
kth moment (exists only for $k < \nu$):

$$E[X^k] = \begin{cases} \dfrac{\Gamma(\frac{1}{2}(k+1))\Gamma(\frac{1}{2}(\nu - k))}{\sqrt{\pi}\,\Gamma(\frac{1}{2}\nu)}\nu^{\frac{1}{2}k}, & k \text{ even}, \\ 0, & k \text{ odd}. \end{cases}$$

$$f(x) = \begin{cases} \dfrac{1}{\beta - \alpha}, & \alpha \leq x \leq \beta \\ 0 & \text{otherwise} \end{cases}; \quad \alpha < \beta.$$

Mean: $E[X] = \frac{1}{2}(\alpha + \beta)$.

34.21   Variance: $\text{var}[X] = \frac{1}{12}(\beta - \alpha)^2$.   *Uniform distribution.*
Moment generating function: $\dfrac{e^{\beta t} - e^{\alpha t}}{t(\beta - \alpha)}$.

Characteristic function: $\dfrac{e^{i\beta t} - e^{i\alpha t}}{it(\beta - \alpha)}$.

$$f(x) = \begin{cases} \beta\lambda^\beta x^{\beta - 1}e^{-(\lambda x)^\beta}, & x > 0 \\ 0, & x \leq 0 \end{cases}; \quad \beta, \lambda > 0.$$

Mean: $E[X] = \dfrac{1}{\lambda}\Gamma\left(1 + \dfrac{1}{\beta}\right)$.   *Weibull distribution. For*
34.22   *$\beta = 1$ we get the expo-*
Variance: $\text{var}[X] = \dfrac{1}{\lambda^2}\left[\Gamma\left(1 + \dfrac{2}{\beta}\right) - \Gamma\left(1 + \dfrac{1}{\beta}\right)^2\right]$.   *nential distribution.*

kth moment: $E[X^k] = \dfrac{1}{\lambda^k}\Gamma(1 + k/\beta)$.

References

See e.g. Evans, Hastings, and Peacock (1993), Johnson, Kotz, and Kemp (1993), Johnson, Kotz, and Balakrishnan (1995), (1997), and Hogg and Craig (1995).

Chapter 35

Method of least squares

Ordinary least squares

The straight line $y = a + bx$ that best fits n data points (x_1, y_1), (x_2, y_2), ..., (x_n, y_n), in the sense that the sum of the squared vertical deviations,

35.1
$$\sum_{i=1}^{n} e_i^2 = \sum_{i=1}^{n} \left[y_i - (a + bx_i) \right]^2,$$

is minimal, is given by the equation

$$y = a + bx \iff y - \bar{y} = b(x - \bar{x}),$$

where

$$b = \frac{\sum_{i=1}^{n}(x_i - \bar{x})(y_i - \bar{y})}{\sum_{i=1}^{n}(x_i - \bar{x})^2}, \quad a = \bar{y} - b\bar{x}.$$

Linear approximation by the *method of least squares*.

$\bar{x} = \frac{1}{n} \sum_{i=1}^{n} x_i,$

$\bar{y} = \frac{1}{n} \sum_{i=1}^{n} y_i.$

35.2
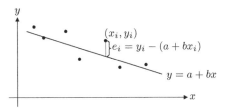

Illustration of the method of least squares with one explanatory variable.

The vertical deviations in (35.1) are $e_i = y_i - y_i^*$, where $y_i^* = a + bx_i$, $i = 1, \ldots, n$. Then **35.3** $\sum_{i=1}^{n} e_i = 0$, and $b = r(s_y^2/s_x^2)$, where r is the correlation coefficient for $(x_1, y_1), \ldots, (x_n, y_n)$. Hence, $b = 0 \iff r = 0$.

$s_x^2 = \frac{1}{n-1} \sum_{i=1}^{n}(x_i - \bar{x})^2,$

$s_y^2 = \frac{1}{n-1} \sum_{i=1}^{n}(y_i - \bar{y})^2.$

In (35.1), the *total variation*, *explained variation*, and *residual variation* in y are defined as

35.4
- Total: $\text{SST} = \sum_{i=1}^{n}(y_i - \bar{y})^2$
- Explained: $\text{SSE} = \sum_{i=1}^{n}(y_i^* - \bar{y}^*)^2$
- Residual: $\text{SSR} = \sum_i e_i^2 = \sum_i (y_i - y_i^*)^2$

Then $\text{SST} = \text{SSE} + \text{SSR}$.

The total variation in y is the sum of the explained and the residual variations.

The correlation coefficient r satisfies

35.5
$$r^2 = \text{SSE/SST},$$

and $100r^2$ is the percentage of explained variation in y.

$r^2 = 1 \Leftrightarrow e_i = 0$ for all i
$\Leftrightarrow y_i = a + bx_i$ (exactly) for all i.

Suppose that the variables x and Y are related by a relation of the form $Y = \alpha + \beta x$, but that observations of Y are subject to random variation. If we observe n pairs (x_i, Y_i) of values of x and Y, $i = 1, \ldots, n$, we can use the formulas in (35.1) to determine least squares estimators $\hat{\alpha}$ and $\hat{\beta}$ of α and β. If we assume that the
35.6 residuals $\varepsilon_i = y_i - \alpha - \beta x_i$ are independently and normally distributed with zero mean and (unknown) variance σ^2, and if the x_i have zero mean, i.e. $\bar{x} = (\sum_i x_i)/n = 0$, then

Linear regression with one explanatory variable. If the x_i do not sum to zero, one can estimate the coefficients in the equation
$Y = \alpha + \beta(x - \bar{x})$
instead.

- the estimators $\hat{\alpha}$ and $\hat{\beta}$ are unbiased,

- $\text{var}(\hat{\alpha}) = \dfrac{\sigma^2}{n}, \quad \text{var}(\hat{\beta}) = \dfrac{\sigma^2}{\sum_i x_i^2}.$

Multiple regression

Given n observations $(x_{i1}, \ldots, x_{ik})$, $i = 1, \ldots, n$, of k quantities $x_1, \ldots, x_k$, and n observations $y_1, \ldots, y_n$ of another quantity y. Define

$$\mathbf{X} = \begin{pmatrix} 1 & x_{11} & x_{12} & \cdots & x_{1k} \\ 1 & x_{21} & x_{22} & \cdots & x_{2k} \\ \vdots & \vdots & \vdots & & \vdots \\ 1 & x_{n1} & x_{n2} & \cdots & x_{nk} \end{pmatrix},$$

35.7
$$\mathbf{y} = \begin{pmatrix} y_1 \\ y_2 \\ \vdots \\ y_n \end{pmatrix}, \quad \mathbf{b} = \begin{pmatrix} b_0 \\ b_1 \\ \vdots \\ b_k \end{pmatrix}.$$

The method of least squares with k explanatory variables.

$\mathbf{X}$ is often called the *design matrix*.

The coefficient vector $\mathbf{b} = (b_0, b_1, \ldots, b_k)'$ of the hyperplane $y = b_0 + b_1 x_1 + \cdots + b_k x_k$ that best fits the given observations in the sense of minimizing the sum

$$(\mathbf{y} - \mathbf{Xb})'(\mathbf{y} - \mathbf{Xb})$$

of the squared deviations, is given by

$$\mathbf{b} = (\mathbf{X'X})^{-1}\mathbf{X'y}.$$

In (35.7), let $y_i^* = b_0 + b_1 x_{i1} + \cdots + b_k x_{ik}$. The sum of the deviations $e_i = y_i - y_i^*$ is then $\sum_{i=1}^n e_i = 0$.

35.8   Define SST, SSE and SSR as in (35.4). Then SST = SSE + SSR and SSR = SST $\cdot (1 - R^2)$, where $R^2 = $ SSE/SST is the *coefficient of determination*. $R = \sqrt{R^2}$ is the *multiple correlation coefficient* between y and the explanatory variables $x_1, \ldots, x_k$.

Definition of the coefficient of determination and the multiple correlation coefficient. $100R^2$ is percentage of explained variation in y.

Suppose that the variables $\mathbf{x} = (x_1, \ldots, x_k)$ and Y are related by an equation of the form $Y = \beta_0 + \beta_1 x_1 + \cdots + \beta_k x_k = (1, \mathbf{x})\boldsymbol{\beta}$, but that observations of Y are subject to random variation. Given n observations $(\mathbf{x}_i, Y_i)$ of values of $\mathbf{x}$ and Y, $i = 1, \ldots, n$, we can use the formulas in (35.7) to determine a least squares estimator $\widehat{\boldsymbol{\beta}} = (\mathbf{X}'\mathbf{X})^{-1}\mathbf{X}'\mathbf{Y}$ of $\boldsymbol{\beta}$. If the resid-

35.9   uals $\varepsilon_i = y_i - (1, \mathbf{x}_i)\boldsymbol{\beta}$ are independently distributed with zero mean and (unknown) variance σ^2, then

- the estimator $\widehat{\boldsymbol{\beta}}$ is unbiased,
- $\mathrm{cov}(\widehat{\boldsymbol{\beta}}) = \sigma^2 (\mathbf{X}'\mathbf{X})^{-1}$,
- $\hat{\sigma}^2 = \dfrac{\hat{\boldsymbol{\varepsilon}}'\hat{\boldsymbol{\varepsilon}}}{n - k - 1} = \dfrac{\sum_i \hat{\varepsilon}_i^2}{n - k - 1}$,
- $\widehat{\mathrm{cov}}(\widehat{\boldsymbol{\beta}}) = \hat{\sigma}^2 (\mathbf{X}'\mathbf{X})^{-1}$.

Multiple regression. $\boldsymbol{\beta} = (\beta_0, \beta_1, \ldots, \beta_k)'$ is the vector of regression coefficients; $\mathbf{x}_i = (x_{i1}, \ldots, x_{ik})$ is the ith observation of $\mathbf{x}$; $\mathbf{Y} = (Y_1, \ldots, Y_n)'$ is the vector of observations of Y; $\hat{\boldsymbol{\varepsilon}} = (\hat{\varepsilon}_1, \ldots, \hat{\varepsilon}_n)' = \mathbf{Y} - \mathbf{X}\widehat{\boldsymbol{\beta}}$; $\mathrm{cov}(\widehat{\boldsymbol{\beta}}) = (\mathrm{cov}(\beta_i, \beta_j))_{ij}$ is the $(n+1) \times (n+1)$ covariance matrix of the vector $\boldsymbol{\beta}$.

If the ε_i are normally distributed, then $\hat{\sigma}^2$ is an unbiased estimator of σ^2, and $\widehat{\mathrm{cov}}(\widehat{\boldsymbol{\beta}})$ is an unbiased estimator of $\mathrm{cov}(\widehat{\boldsymbol{\beta}})$.

References

See e.g. Hogg and Craig (1995) or Rice (1995).

References

Anderson, I.: *Combinatorics of Finite Sets*, Clarendon Press (1987).

Barro, R. J. and X. Sala-i-Martin: *Economic Growth*, McGraw-Hill (1995).

Bartle, R. G.: *Introduction to Real Analysis*, John Wiley & Sons (1982).

Beavis, B. and I. M. Dobbs: *Optimization and Stability Theory for Economic Analysis*, Cambridge University Press (1990).

Bellman, R.: *Dynamic Programming*, Princeton University Press (1957).

Bhagwati, J .N., A. Panagariya and T. N. Srinivasan: *Lectures on International Trade.* 2nd ed., MIT Press (1998).

Black, F. and M. Scholes: "The pricing of options and corporate liabilities", *Journal of Political Economy*, Vol. 81, 637–654 (1973).

Blackorby, C. and R. R. Russell: "Will the real elasticity of substitution please stand up? (A comparison of the Allen/Uzawa and Morishima elasticities)", *American Economic Review*, Vol. 79, 882–888 (1989).

Blanchard, O. and S. Fischer: *Lectures on Macroeconomics*, MIT Press (1989).

Braun, M.: *Differential Equations and their Applications*, 4th ed., Springer (1993).

Burmeister, E. and A. R. Dobell: *Mathematical Theories of Economic Growth*, Macmillan (1970).

Charalambides, C. A.: *Enumerative Combinatorics*, Chapman & Hall/CRC (2002).

Christensen, L. R., D. Jorgenson and L. J. Lau: "Transcendental logarithmic utility functions", *American Economic Review*, Vol. 65, 367–383 (1975).

Clarke, F. H.: *Optimization and Nonsmooth Analysis.* John Wiley & Sons (1983).

Deaton, A. and J. Muellbauer: *Economics and Consumer Behaviour*, Cambridge University Press (1980).

Dhrymes, P. J.: *Mathematics for Econometrics*, Springer (1978).

Dixit, A. K.: *Optimization in Economic Theory*, 2nd ed., Oxford University Press (1990).

Edwards, C. H. and D. E. Penney: *Calculus with Analytic Geometry*, 5th ed., Prentice-Hall (1998).

Ellickson, B.: *Competitive Equilibrium. Theory and Applications*, Cambridge University Press (1993).

Evans, M., N. Hastings, and B. Peacock: *Statistical Distributions*, 2nd ed., John Wiley & Sons (1993).

Faddeeva, V. N.: *Computational Methods of Linear Algebra*, Dover Publications, Inc. (1959).

Farebrother, R. W.: "Simplified Samuelson conditions for cubic and quartic equations", *The Manchester School of Economic and Social Studies*, Vol. 41, 396–400 (1973).

Feichtinger, G. and R. F. Hartl: *Optimale Kontrolle Ökonomischer Prozesse*, Walter de Gruyter (1986).

Fleming, W. H. and R. W. Rishel: *Deterministic and Stochastic Optimal Control*, Springer (1975).

Førsund, F.: "The homothetic production function", *The Swedish Journal of Economics*, Vol. 77, 234–244 (1975).

Fraleigh, J. B. and R. A. Beauregard: *Linear Algebra*, 3rd ed., Addison-Wesley (1995).

Friedman, J. W.: *Game Theory with Applications to Economics*, Oxford University Press (1986).

Fudenberg, D. and J. Tirole: *Game Theory*, MIT Press (1991).

Fuss, M. and D. McFadden (eds.): *Production Economics: A Dual Approach to Theory and Applications*, Vol. I, North-Holland (1978).

Gandolfo, G.: *Economic Dynamics*, 3rd ed., Springer (1996).

Gantmacher, F. R.: *The Theory of Matrices*, Chelsea Publishing Co. (1959). Reprinted by the American Mathematical Society, AMS Chelsea Publishing (1998).

Gass, S. I.: *Linear Programming. Methods and Applications*, 5th ed., McGraw-Hill (1994).

Gibbons, R.: *A Primer in Game Theory*, Harvester and Wheatsheaf (1992).

Goldberg, S.: *Introduction to Difference Equations*, John Wiley & Sons (1961).

Graham, R. L., D. E. Knuth and O. Patashnik: *Concrete Mathematics*, Addison-Wesley (1989).

Griffiths, W. E., R. Carter Hill and G. G. Judge: *Learning and Practicing Econometrics*, John Wiley & Sons (1993).

Hadar, J. and W. R. Russell: "Rules for ordering uncertain prospects", *American Economic Review*, Vol. 59, 25–34 (1969).

Halmos, P. R.: *Naive Set Theory*, Springer (1974).

Hardy, G. H., J. E. Littlewood, and G. Pólya: *Inequalities*, Cambridge University Press (1952).

Hartman, P.: *Ordinary Differential Equations*, Birkhäuser (1982).

Haug, E. G.: *The Complete Guide to Option Pricing Formulas*, McGraw-Hill (1997).

Hildebrand, F. B.: *Finite-Difference Equations and Simulations*, Prentice-Hall (1968).

Hildenbrand, W.: *Core and Equilibria of a Large Economy*, Princeton University Press (1974).

Hildenbrand, W. and A. P. Kirman: *Introduction to Equilibrium Analysis*, North-Holland (1976).

Hogg, R. V. and A. T. Craig: *Introduction to Mathematical Statistics*, 5th ed., Prentice-Hall (1995).

Horn, R. A. and C. R. Johnson: *Matrix Analysis*, Cambridge University Press (1985).

Huang, Chi-fu and R. H. Litzenberger: *Foundations for Financial Economics*, North-Holland (1988).

Intriligator, M. D.: *Mathematical Optimization and Economic Theory*, Prentice-Hall (1971).

Johansen, L.: *Production Functions*, North-Holland (1972).

Johnson, N. L., S. Kotz, and S. Kemp: *Univariate Discrete Distributions*, 2nd ed., John Wiley & Sons (1993).

Johnson, N. L., S. Kotz, and N. Balakrishnan: *Continuous Univariate Discrete Distributions*, John Wiley & Sons (1995).

Johnson, N. L., S. Kotz, and N. Balakrishnan: *Discrete Multivariate Distributions*, John Wiley & Sons (1997).

Johnson, R. A. and G. K. Bhattacharyya: *Statistics: Principles and Methods*, 3rd ed., John Wiley & Sons (1996).

Kamien, M. I. and N. I. Schwartz: *Dynamic Optimization: the Calculus of Variations and Optimal Control in Economics and Management*, 2nd ed., North-Holland (1991).

Karatzas, I. and S. E. Shreve: *Brownian Motion and Stochastic Calculus*, 2nd ed., Springer (1991).

Kolmogorov, A. N. and S. V. Fomin: *Introductory Real Analysis*, Dover Publications (1975).

Kreps, D. M.: *A Course in Microeconomic Theory*, Princeton University Press (1990).

Lang, S.: *Linear Algebra*, 3rd ed., Springer (1987).

Larsen, R. J. and M. L. Marx: *An Introduction to Mathematical Statistics and its Applications*, Prentice-Hall (1986).

Léonard, D. and N. Van Long: *Optimal Control Theory and Static Optimization in Economics*, Cambridge University Press (1992).

Luenberger, D. G.: *Introduction to Linear and Nonlinear Programming*, 2nd ed., Addison-Wesley (1984).

Lütkepohl, H.: *Handbook of Matrices*, John Wiley & Sons (1996).

Magnus, J. R. and H. Neudecker: *Matrix Differential Calculus with Applications in Statistics and Econometrics*, John Wiley & Sons (1988).

Marsden, J. E. and M. J. Hoffman: *Elementary Classical Analysis*, 2nd ed., W. H. Freeman and Company (1993).

Mas-Colell, A., M. D. Whinston, and J. R. Green: *Microeconomic Theory*, Oxford University Press (1995).

Merton, R. C.: "Theory of rational option pricing", *Bell Journal of Economics and Management Science*, Vol. 4, 141–183 (1973).

Nikaido, H.: *Convex Structures and Economic Theory*, Academic Press (1968).

Nikaido, H.: *Introduction to Sets and Mappings in Modern Economics*, North-Holland (1970).

Norstrøm, C. J.: "A sufficient condition for a unique nonnegative internal rate of return", *Journal of Financial and Quantitative Analysis*, Vol. 7, 1835–1839 (1972).

Øksendal, B.: *Stochastic Differential Equations, an Introduction with Applications*, 6th ed., Springer (2003).

Parthasarathy, T.: *On Global Univalence Theorems*. Lecture Notes in Mathematics. No. 977, Springer (1983).

Phlips, L.: *Applied Consumption Analysis*, North-Holland (1983).

Rice, J: *Mathematical Statistics and Data Analysis*, 2nd ed., Duxburry Press (1995).

Rockafellar, T.: *Convex Analysis*, Princeton University Press (1970).

Rothschild, M. and J. Stiglitz: "Increasing risk: (1) A definition", *Journal of Economic Theory*, Vol. 2, 225–243 (1970).

Royden, H. L.: *Real Analysis*, 3rd ed., Macmillan (1968).

Rudin, W.: *Principles of Mathematical Analysis*, 2nd ed., McGraw-Hill (1982).

Scarf, H. (with the collaboration of T. Hansen): *The Computation of Economic Equilibria*. Cowles Foundation Monograph, 24, Yale University Press (1973).

Seierstad, A. and K. Sydsæter: *Optimal Control Theory with Economic Applications*, North-Holland (1987).

Sharpe, W. F.: "Capital asset prices: A theory of market equilibrium under conditions of risk", *Journal of Finance*, Vol. 19, 425–442 (1964).

Shephard, R. W.: *Cost and Production Functions*, Princeton University Press (1970).

Silberberg, E.: *The Structure of Economics. A Mathematical Analysis*, 2nd ed., McGraw-Hill (1990).

Simon, C. P. and L. Blume: *Mathematics for Economists*, Norton (1994).

Sneddon, I. N.: *Elements of Partial Differential Equations*, McGraw-Hill (1957).

Stokey, N. L. and R. E. Lucas, with E. C. Prescott: *Recursive Methods in Economic Dynamics*, Harvard University Press (1989).

Sundaram, R. K.: *A First Course in Optimization Theory*, Cambridge University Press (1996).

Sydsæter, K. and P. J. Hammond: *Essential Mathematics for Economic Analysis*, FT Prentice Hall (2005).

Sydsæter, K., P. J. Hammond, A. Seierstad, and Arne Strøm: *Further Mathematics for Economic Analysis*, 2nd ed., FT Prentice Hall (2005).

Takayama, A.: *Mathematical Economics*, 2nd ed., Cambridge University Press (1985).

Topkis, Donald M.: *Supermodularity and Complementarity*, Princeton University Press (1998).

Turnbull, H. W.: *Theory of Equations*, 5th ed., Oliver & Boyd (1952).

Varian, H.: *Microeconomic Analysis*, 3rd ed., Norton (1992).

Weibull, J. W.: *Evolutionary Game Theory*, MIT Press (1995).

Zachmanoglou, E. C. and D. W. Thoe: *Introduction to Partial Differential Equations with Applications*, Dover Publications (1986).

Index

$Q_1(x_1, x_2) = x_1^2 + x_2^2$ $Q_2(x_1, x_2) = -x_1^2 - x_2^2$

PD   ND

$$A = \begin{pmatrix} a_{11} & a_{12} & a_{13} \\ a_{21} & a_{22} & a_{23} \\ a_{31} & a_{32} & a_{33} \end{pmatrix}$$

Third order principal minor: $\det(A)$. There are three second order principal minor:

① $\begin{vmatrix} a_{11} & a_{12} \\ a_{21} & a_{22} \end{vmatrix}$ ② $\begin{vmatrix} a_{11} & a_{13} \\ a_{31} & a_{33} \end{vmatrix}$ ③ $\begin{vmatrix} a_{22} & a_{23} \\ a_{32} & a_{33} \end{vmatrix}$

There are 3 first order principal minor: $|A_{11}| \, |A_{22}| \, |A_{33}|$

Definiteness of Quadratic Forms

Each quadratic form Q can be represented by a symmetric matrix A so that
$Q(x) = x^T \cdot A \cdot x$.

Matrix representation of quadratic forms:

$$a_{11} x_1^2 + a_{12} x_1 x_2 + a_{22} x_2^2 = (x_1 \ x_2) \begin{pmatrix} a_{11} & \frac{1}{2}a_{12} \\ \frac{1}{2}a_{12} & a_{22} \end{pmatrix} \begin{pmatrix} x_1 \\ x_2 \end{pmatrix}$$

$$Q(x_1 x_2 x_3) = a_{11}x_1^2 + a_{12}x_1 x_2 + a_{13}x_1 x_3 + a_{22}x_2^2 + a_{23}x_2 x_3 + a_{33}x_3^2$$

$$= (x_1 \ x_2 \ x_3) \begin{pmatrix} a_{11} & \frac{1}{2}a_{12} & \frac{1}{2}a_{13} \\ \frac{1}{2}a_{21} & a_{22} & \frac{1}{2}a_{23} \\ \frac{1}{2}a_{31} & \frac{1}{2}a_{32} & a_{33} \end{pmatrix} \begin{pmatrix} x_1 \\ x_2 \\ x_3 \end{pmatrix}$$

General quadratic form of one variable is $y = ax^2$. If $a > 0$, then ax^2 is always ≥ 0 & equals 0 only when $x = 0$. This is **PD**. $x = 0$ is a global minimizer.
If $a < 0$, then $ax^2 \leq 0$ and equals 0 only when $x = 0$. This is **ND** & $x =$ global maximizer.

$Q_1(x_1, x_2) = x_1^2 + x_2^2$ is always > 0 @ $(x_1, x_2) \neq 0$. So, **PD**. $Q_2(x_1, x_2) = -x_1^2 - x_2^2$ which are strictly neg. @ the origin are **ND**. $Q_3(x_1, x_2) = x_1^2 - x_2^2$ are indefinite. (Saddle Shape)
A quadratic form, such as $Q_4(x_1, x_2) = (x_1 + x_2)^2 = x_1^2 + 2x_1 x_2 + x_2^2$, which is always greater than or equal to $\emptyset$, but may $= \emptyset$ @ some nonzero x's is **PSD**. (Folded paper up)
a quadratic form, like $Q_5(x_1, x_2) = -(x_1 + x_2)^2$ which is never positive but can be zero @ points other than the origin is **NSD** (Folded paper down)

Definite symmetric matrix
A) PD if $x^T A x > 0$ for all $x \neq 0$ in $\mathbb{R}^n$.
B) PSD if $x^T A x \geq 0$ for all $x \neq 0$ in $\mathbb{R}^n$.
C) ND if $x^T A x < 0$ for all $x \neq 0$ in $\mathbb{R}^n$.
D) NSD if $x^T A x \leq 0$ for all $x \neq 0$ in $\mathbb{R}^n$ and
e) Indefinite if $x^T A x > 0$ for some x in $\mathbb{R}^n$ and < 0 for some other x in $\mathbb{R}^n$.

If PD $\Rightarrow$ then PSD. If ND, then NSD.

Theorem 16.1 (works only if No constraints)
Let A be an $n \times n$ symmetric matrix.
a) A is PD iff all its n leading principal minors are strictly positive.
b) A is ND iff its n leading principal minors alternate signs: $D_1 < 0 \ D_2 > 0 \ D_3 < 0$
c) If some kth order leading principal minor of A is nonzero, but does not follow pattern, it is indefinite.
d) If some leading principal minor $= 0$, but nonzero minors follow pattern, the matrix A is NOT definite & may or may not be semi definite. Then, need Theorem 16.2.

Theorem 16.2
Matrix A is PSD iff every principal minor of A is ≥ 0; A is NSD iff every principal minor of odd order is ≤ 0 & every principal minor of even order is $\geq$

EX 16.3 For 4×4 symmetric matrix:
A) If $|A_1| > 0, |A_2| > 0, |A_3| > 0, |A_4| > 0$, then **PD**. D) If $|A_1| < 0, |A_2| < 0, |A_3| < 0, |A_4| < 0$, Indef
B) If $|A_1| < 0, |A_2| > 0, |A_3| < 0, |A_4| > 0$, then **ND**. E) If $|A_1| = 0, |A_2| < 0, |A_3| > 0, |A_4| = 0$, Indef.
C) If $|A_1| > 0, |A_2| > 0, |A_3| = 0, |A_4| < 0$, then dndef. F) If $|A_1| > 0, |A_2| = 0, |A_3| > 0, |A_4| > 0$, Not def.
Need to ck all other minors. If none neg. then A is PSD. Otherwise, indefinite.

Definition of optimality

$X, Q(X, X_2) = X_1^2 - X_2^2$ is indefinite. But impose constraint of $X_2 = 0$. Then,

$Q(X, 0) = X_1^2$ and there is strict global min @ $X_1 = 0$. & Q is PD w/ constraint set

$\{X_2 = 0\}$. If we impose constraint $X_1 = 0$ then $X_2 = 0$ is a global max of $Q(0, X_2) =$

X_2^2 and Q is ND for $\{X_1 = 0\}$

n the line $X_1 - 2X_2 = 0$, $Q(2X_2, X_2) = (2X_2)^2 - X_2^2 = 3X_2^2$ is PD,

Constrained optimization

Bordered Hessian Example:

$Q(X, X_2) = aX_1^2 + 2bX_1 X_2 + cX_2^2$

Constraint: $AX_1 + BX_2 = 0$

o determine definiteness of quadratic
orm of "n" variables, $Q(x) = x^T A x$,
hen restricted by a constraint given
by "m" linear equations $Bx = 0$, construct
he $(n+m) \times (n+m)$ symmetric matrix
I by bordering the matrix A above & to the
eft by the coefficients B of the linear
nstraints: $H = \begin{pmatrix} 0 & B \\ B^T & A \end{pmatrix}$

eck the sign of the last n-m leading
incipal minors of H, start w/ determinant
elf.

If det H has same sign as $(-1)^n$ & if the
ast n-m leading principal minors
lternate signs, Q is ND on constraint
et $Bx = 0$, and $x = 0$ is a strict global
max of Q on this constraint set.

If det H and the last n-m leading principal
minors all have the same sign as $(-1)^m$,
then Q is PD on constrained set $Bx = 0$, &
$X = 0$ is a strict global min.

If both A) + B) are violated by nonzero
leading principal minors, then Q is indefinite.
on the constraint $Bx = 0$ & $X = 0$ is not either
max or min of Q on this constraint set.

$\mathcal{L}(x, \gamma) = \begin{pmatrix} 0 & -g_1 & -g_2 \\ -g_1 & f_{11} - \gamma g_{11} & f_{12} - \gamma g_{12} \\ -g_2 & f_{12} - \gamma g_{12} & f_{22} - \gamma g_{22} \end{pmatrix}$

/example

$\mathcal{L}(x, \gamma) = \begin{pmatrix} 0 & -4 & -2 \\ -4 & 0 & 1 \\ -2 & 1 & 0 \end{pmatrix}$

Bordered Hessian

$\bar{H} = \begin{pmatrix} 0 & g_1 & g_2 \\ g_1 & h_{11} & h_{12} \\ g_2 & h_{12} & h_{22} \end{pmatrix}$

$\det \begin{pmatrix} 0 & A & B \\ A & a & b \\ B & b & c \end{pmatrix}$

PD iff det is negative.
ND iff det is positive.

EX. max
$(X_1, X_2) \in \mathbb{R}^2$ $X_1 X_2 + 2X_1$
such that $4X_1 + 2X_2 = 60$

$\mathcal{L}(x, \gamma) = X_1 X_2 + 2X_1 - \gamma(4X_1 + 2X_2 - 60)$

$\frac{\partial \mathcal{L}}{\partial X_1} = X_2 + 2 - 4\gamma; \quad X_2 + 2 = 4\gamma$
$= 0$

$\frac{\partial \mathcal{L}}{\partial X_2} = X_1 - 2\gamma; \quad X_1 = 2\gamma$
$= 0$

$\frac{\partial \mathcal{L}}{\partial \gamma} = 4X_1 + 2X_2 - 60; \quad 4X_1 + 2X_2 = 60$
$= 0$

using 1 & 2: $\frac{X_2 + 2}{X_1} = \frac{4\gamma}{2\gamma}$

$\frac{X_2 + 2}{X_1} = 2; \quad X_2 + 2 = 2X_1$
$\quad X_2 = 2X_1 - 2$

Then,
$4X_1 + 2(2X_1 - 2) = 60$
$X_1^* = 8$

$4(8) + 2X_2 = 60$
$X_2^* = 14$

Critical Points:
(X_1^*, X_2^*, γ^*)
$(8, 14, 4)$

$X_1 = 2\gamma$
$8 = 2\gamma$
$\gamma = 4$

2nd order on $\mathcal{L}$:
$\frac{\partial^2 \mathcal{L}}{(\partial X_1)^2} = 0 \quad \frac{\partial^2 \mathcal{L}}{(\partial X_2)^2} = 0 \quad \frac{\partial^2 \mathcal{L}}{\partial X_1 \partial X_2} = 1$

$\frac{\partial^2 \mathcal{L}}{\partial X_1 \partial \gamma} = -4 \quad \frac{\partial^2 \mathcal{L}}{\partial X_2 \partial X_1} = 1 \quad \frac{\partial^2 \mathcal{L}}{\partial X_2 \partial \gamma} = -2$

$\frac{\partial^2 \mathcal{L}}{(\partial \gamma)^2} = 0 \quad \frac{\partial^2 \mathcal{L}}{\partial \gamma \partial X_1} = -4 \quad \frac{\partial^2 \mathcal{L}}{\partial \gamma \partial X_2} = -2$

$D^2 \mathcal{L}(x, \gamma) = \begin{matrix} 0 & -4 & -2 \\ -4 & 0 & 1 \\ -2 & 1 & 0 \end{matrix}$

$|D^2 \mathcal{L}(x, \gamma)| = +4(0 - (-2)) + (-2)(-4 - 0)$
$= +8 + 8 = 16 > 0$

$D^2 \mathcal{L}(x, \gamma) = \begin{pmatrix} 0 & -4 & -2 \\ -4 & 0 & 1 \\ -2 & 1 & 0 \end{pmatrix}$

$D_1 = 0 \quad D_2 = -16 \quad D_3 = 16$

continued

Bordered
Hessian
Determinant

equality constraint w/ Lagrangian analysis

In general. max $f(x)$ where $x \in \mathbb{R}^n$ such that $g(x) = K$

$\mathcal{L}(M, \tau) = f(x) - \tau(g(x) - K)$

$\mathcal{L} : \mathbb{R}^{N+1} \to \mathbb{R}$ |The trick| : If $(g(x) - K) = 0$, then A) $g(x) = K$ is satisfied

B) $\mathcal{L}(M, \tau) = f(x)$

Ex, in abstract

① A) $\mathcal{L}_1 = f_1 - \tau g_1$
$= 0 \iff f_1 = \tau g_1$ ⎫ FOC

B) $\mathcal{L}_2 = f_2 - \tau g_2$
$= 0 \iff f_2 = \tau g_2$ ⎬

C) $\mathcal{L}_\tau = -(g(x) - K)$
$= 0 \iff g(x) = K$ ⎭

$D^2\mathcal{L} = \begin{bmatrix} 0 & -g_1 & -g_2 \\ -g_1 & f_{11} - \tau g_{11} & f_{12} - \tau g_{12} \\ -g_2 & f_{12} - \tau g_{12} & f_{22} - \tau g_{22} \end{bmatrix}$

② using A ≗ B:
$\dfrac{f_1}{f_2} = \dfrac{\tau g_1}{\tau g_2}$

$\dfrac{f_1}{f_2} = \dfrac{g_1}{g_2}$

③ Find critical points
(x_1^*, x_2^*, τ)
and evaluate

④ soc!
$\mathcal{L}_1 = f_1 - \tau g_1$
$\mathcal{L}_{11} = f_{11} - \tau g_{11}$
$\mathcal{L}_{12} = f_{12} - \tau g_{12}$
$\mathcal{L}_{1\tau} = -g_1$
$\mathcal{L}_{12} = f_{12} - \tau g_{12}$
$\mathcal{L}_{22} = f_{22} - \tau g_{22}$
$\mathcal{L}_{2\tau} = -g_2$
$\mathcal{L}_{1\tau} = -g_1$
$\mathcal{L}_{2\tau} = -g_2$
$\mathcal{L}_{\tau\tau} = 0 \,(= -g_\tau)$

Suppose only concerned w/ $h(x) = f(x) - \tau(g(x))$ for variation in x?

FOC: $h_1 = f_1 - \tau g_1 = 0$
$h_2 = f_2 - \tau g_2 = 0$

SO (sufficient) Condition:
$D_2 h = \begin{pmatrix} h_{11} & h_{12} \\ h_{12} & h_{22} \end{pmatrix} = \begin{pmatrix} f_{11} - \tau g_{11} & f_{12} - \tau g_{12} \\ f_{12} - \tau g_{12} & f_{22} - \tau g_{22} \end{pmatrix}$

Is ND
$\iff h_{11} < 0$ and $|D^2 h| > 0$
$\iff [dx_1, dx_2] [D^2 h] \begin{bmatrix} dx_1 \\ dx_2 \end{bmatrix} < 0 \; \forall \; \begin{bmatrix} dx_1 \\ dx_2 \end{bmatrix}$

Looking @ $h(x) = f(x) - \tau g(x)$ $h \in \mathbb{R}^2 \to \mathbb{R}$

$D^2 h = \begin{pmatrix} h_{11} & h_{12} \\ h_{12} & h_{22} \end{pmatrix}$

want this to be neg. definite
$\iff h_{11} < 0 \quad |D^2 h| > 0$
$\iff \underset{1 \times 2}{[dx_1, dx_2]} \; \underset{1 \times 1}{[D^2 h]} \; \underset{2 \times 1}{\begin{bmatrix} dx_1 \\ dx_2 \end{bmatrix}} < 0$

$h_{11}(dx_1)^2 + 2h_{12} dx_1 dx_2 + h_{22}(dx_2)^2 < 0$

Have a constraint $\to g(x_1, x_2) = K$

implicit differentiation

$\dfrac{\partial g}{\partial x_1}(x) dx_1 + \dfrac{\partial g}{\partial x_1}(x) dx_2 = 0$

$g_1 dx_1 + g_2 dx_2 = 0$

$dx_2 = \dfrac{-g_1}{g_2} dx_1$

$h_{11}(dx_1)^2 + 2h_{12} dx_1 \left(-\dfrac{g_1}{g_2} dx_1\right) + h_{22}\left(-\dfrac{g_1}{g_2} dx_1\right)^2 < 0$

$\underbrace{\dfrac{(dx_1)^2}{(g_2)^2}}_{\text{Positive}} \underbrace{\left(h_{11}(g_2)^2 - 2h_{12} g_1 g_2 + h_{22}(g_1)^2\right)}_{\text{negative}} < 0$

$\iff h_{11}(g_2)^2 - 2h_{12}g_1 g_2 + h_{22}(g_1)^2$
$\iff |\overline{H}| > 0$
now look @ $|\overline{H}| = \begin{vmatrix} 0 & g_1 & g_2 \\ g_1 & h_{11} & h_{12} \\ g_2 & h_{12} & h_{22} \end{vmatrix}$

$= -g_1(g_1 h_{22} - g_2 h_{12}) + g_2(g_1 h_{12} - g_2 h_{12} \cdots$
$= -h_{22}(g_1)^2 + h_{12} g_1 g_2 + h_{12} g_1 g_2 \cdots$
$= -h_{11}(g_2)^2 + 2h_{12} g_1 g_2 - h_{22}(g_1)$

Degrees of Freedom!

Defn: # of parameters of a system that may vary independently.

	Degrees of Freedom
a) max $f(x)$ $\;x \in \mathbb{R}$	1
b) max $f(x)$ $\;x \in \mathbb{R}^2$	2
c) max $f(x)$ $\;x \in \mathbb{R}^3$	3
d) max $f(x)$ $\;x \in \mathbb{R}^2$ w/ constraint $g(x) = K$	1
e) max $f(x)$ $\;x \in \mathbb{R}^3$ w/ constraint $g_a(x) = Ka$ $g_b(x) = Kb$	1 (have 2 constrai

continued →

CPSIA information can be obtained at www.ICGtesting.com
Printed in the USA
LVOW01*2037270713

344969LV00007BA/336/P

9 783540 260882

we max $f(x)$ s.t. $g_i(x) = a_i$
$x \in \mathbb{R}^n$

$$g_i(x) = a_i$$
$$h_1(x) \leq b_1$$
$$h_j(x) \leq b_j$$

Lagrange:
$$\mathcal{L}(x, \tau) = f(x)$$
$$\sum_{i=1}^{I} \tau_{Gi}(g_i(x) - a_i)$$
$$\sum_{j=1}^{J} \tau_{Hj}(h_j(x) - b_j)$$

if this is nonzero#
then τ_{Hj} needs to
be zero.

k: (called "Kuhn-Tucker conditions")
$$\frac{\partial \mathcal{L}}{\partial x_k}(x, \tau) = \frac{\partial f}{\partial x_k}(x) - \sum_{i=1}^{I} \tau_{Gi}\left(\frac{\partial g_i(x)}{\partial x_k}\right)$$
$$- \sum_{j=1}^{J} \tau_{Hj}\left(\frac{\partial h_j}{\partial x_2}(x)\right) = 0 \ \forall k = 1 \ldots N$$

all original constraints
$g_i(x) = a_i$, $h_1(x) = b_1$
$g_i(x) = a_i$ $h_j(x) = b_j$
Inequalities on the τ_H multiplier:
$\tau_{Hj} \geq 0 \ \forall j = 1 \ldots J$
Complementary slackness condition
$\tau_{Hj}(h_j(x) - b_j) = 0 \ \forall j = 1 \ldots J$

slack means
they 2 don't
equal.

$$\frac{dy}{dt} = f(y)$$

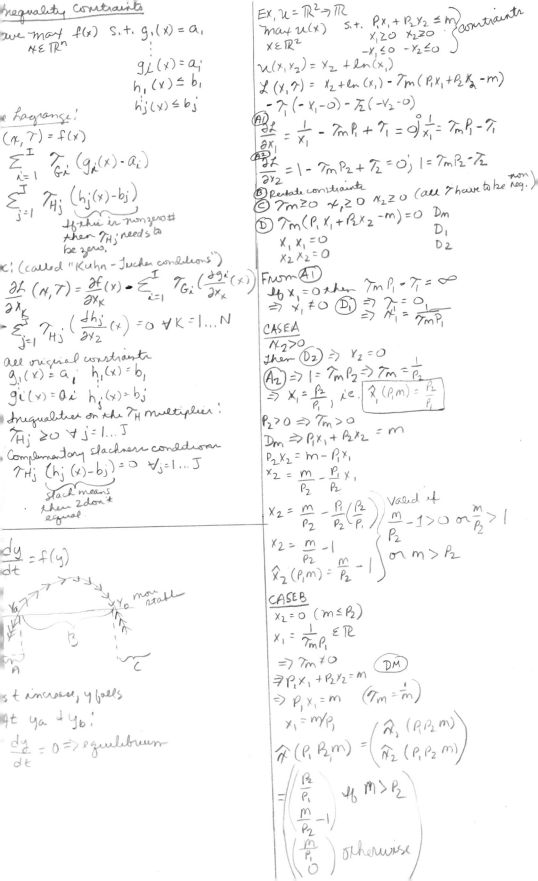

s t increase, y falls
At y_a & y_b:
$\frac{dy}{dt} = 0 \Rightarrow$ equilibrium

EX, $u = \mathbb{R}^2 \to \mathbb{R}$
max $u(x)$ s.t. $P_1 x_1 + P_2 x_2 \leq m$ } constraints
$x \in \mathbb{R}^2$ $\quad x_1 \geq 0 \ x_2 \geq 0$
$\quad\quad\quad -x_1 \leq 0 \ -x_2 \leq 0$

$u(x_1, x_2) = x_2 + \ln(x_1)$
$\mathcal{L}(x, \tau) = x_2 + \ln(x_1) - \tau_m(P_1 x_1 + P_2 x_2 - m)$
$\quad - \tau_1(-x_1 - 0) - \tau_2(-x_2 - 0)$

(A1) $\frac{\partial \mathcal{L}}{\partial x_1} = \frac{1}{x_1} - \tau_m P_1 + \tau_1 = 0; \frac{1}{x_1} = \tau_m P_1 - \tau_1$

(A2) $\frac{\partial \mathcal{L}}{\partial x_2} = 1 - \tau_m P_2 + \tau_2 = 0; \ 1 = \tau_m P_2 - \tau_2$

(B) Restate constraints
(C) $\tau_m \geq 0 \ x_1 \geq 0 \ x_2 \geq 0$ (all τ have to be non-neg.)
(D) $\tau_m(P_1 x_1 + P_2 x_2 - m) = 0$ Dm
$\quad x_1 x_1 = 0$ D1
$\quad x_2 x_2 = 0$ D2

From (A1)
If $x_1 = 0$ then $\tau_m P_1 - \tau_1 = \infty$
$\Rightarrow x_1 \neq 0$ (D1) $\Rightarrow \tau_1 = 0$
$\quad\quad\quad\quad \Rightarrow \tau_1 = \frac{1}{\tau_m P_1}$

CASE A
$x_2 > 0$
then (D2) $\Rightarrow x_2 = 0$
(A2) $\Rightarrow 1 = \tau_m P_2 \Rightarrow \tau_m = \frac{1}{P_2}$
$\Rightarrow x_1 = \frac{P_2}{P_1}$, i.e. $\hat{x}_1(P, m) = \frac{P_2}{P_1}$

$P_2 > 0 \Rightarrow \tau_m > 0$
Dm $\Rightarrow P_1 x_1 + P_2 x_2 = m$
$P_2 x_2 = m - P_1 x_1$
$x_2 = \frac{m}{P_2} - \frac{P_1 x_1}{P_2}$

$x_2 = \frac{m}{P_2} - \frac{P_1}{P_2}\left(\frac{P_2}{P_1}\right)$ $\left.\begin{array}{l}\text{Valid if}\\ \frac{m}{P_2} - 1 > 0 \text{ or } \frac{m}{P_2} > 1\\ \text{or } m > P_2\end{array}\right.$

$x_2 = \frac{m}{P_2} - 1$
$\hat{x}_2(P, m) = \frac{m}{P_2} - 1$

CASE B
$x_2 = 0$ $(m \leq P_2)$
$x_1 = \frac{1}{\tau_m P_1} \in \mathbb{R}$
$\Rightarrow \tau_m \neq 0$ (DM)
$\Rightarrow P_1 x_1 + P_2 x_2 = m$
$\Rightarrow P_1 x_1 = m$ $(\tau_m = \frac{1}{m})$
$x_1 = m/P_1$

$$\hat{x}(P_1, P_2, m) = \begin{pmatrix} \hat{x}_1(P_1, P_2, m) \\ \hat{x}_2(P_1, P_2, m) \end{pmatrix}$$

$$= \begin{pmatrix} \frac{P_2}{P_1} \\ \frac{m}{P_2} - 1 \end{pmatrix} \text{ if } m > P_2$$

$$\begin{pmatrix} \frac{m}{P_1} \\ 0 \end{pmatrix} \text{ otherwise}$$

Characteristic Values & Vectors

$A = \begin{pmatrix} 1 & 0 \\ 0 & 2 \end{pmatrix}$

$|A - rI| = \left| \begin{pmatrix} 1 & 0 \\ 0 & 2 \end{pmatrix} - \begin{pmatrix} r & 0 \\ 0 & r \end{pmatrix} \right| = \left| \begin{matrix} 1-r & 0 \\ 0 & 2-r \end{matrix} \right| = (1-r)(2-r) = 0$

Char. values are 1 & 2.

Derivatives & limits

$\dfrac{\Delta y}{\Delta x} = \dfrac{f(x^0 + \Delta x) - f(x^0)}{\Delta x}$

as $\Delta x \to 0$

$f(x) = x^2$

EX. $\lim\limits_{x \to 2} f(x) = f(2) = 2(2) = 4$

EX $\lim\limits_{x \to +0} \dfrac{2x+5}{x+1} = \dfrac{2x+2+3}{x+1} = \dfrac{2(x+1)+3}{x+1} = 2 + \dfrac{3}{x+1} = 2$

EX. $f(x) = \begin{cases} 2x & \text{if } x \le 4 \\ 2x + 4 & \text{if } x > 4 \end{cases}$ Limit Does not exist

$\lim\limits_{x \to 4^-} = 8$

$\lim\limits_{x \to 4^+} = 12$

EX, $f(x) = x^2$

$f'(x) = \lim\limits_{\Delta x \to 0} \dfrac{f(x + \Delta x) - f(x)}{\Delta x}$

$= \lim\limits_{\Delta x \to 0} \dfrac{(x + \Delta x)^2 - (x)^2}{\Delta x}$

$= \lim\limits_{\Delta x \to 0} 2x + \Delta x = 2x$

Discontinuous

1) $\lim\limits_{x \to x^0} b = b$

2) $\lim\limits_{x \to x^0} bx = bx^0$

3) $\lim\limits_{x \to x^0} (x)^n = (x^0)^n$

$n \ge 1$ or $n \ge 0$ and $x_0 \ge 0$

avoiding this

Inverse Function Rule

EX. $f(x) = x^3 \quad g(x) = (x)^{1/3}$

$f'(x) = \dfrac{1}{g'(x^3)} = \dfrac{1}{\frac{1}{3}(x^3)^{-2/3}}$

$= 3(x^3)^{2/3} = 3x^2$

$\exp(x+y) = \exp(x)\,\exp(y)$

$b = \ln(a)$

$a = \exp(b) = e^b$

$b = \ln(a) \Longleftrightarrow a = e^b$

$\ln(xy) = \ln(x) + \ln(y)$

$\ln(x^\alpha) = \alpha \ln(x)$